Josiah Conder

The floral art of Japan

Being a second and revised edition of the flowers of Japan and the art of floral arrangement

Josiah Conder

The floral art of Japan
Being a second and revised edition of the flowers of Japan and the art of floral arrangement

ISBN/EAN: 9783337174927

Printed in Europe, USA, Canada, Australia, Japan

Cover: Foto ©berggeist007 / pixelio.de

More available books at **www.hansebooks.com**

THE

FLORAL ART OF JAPAN:

BEING

A SECOND AND REVISED EDITION

OF

THE FLOWERS OF JAPAN

AND THE

ART OF FLORAL ARRANGEMENT.

BY

JOSIAH CONDER, F.R.I.B.A.,

WITH ILLUSTRATIONS BY JAPANESE ARTISTS.

TOKIO

PRINTED BY THE SHUYEI SHA; PUBLISHED BY KELLY AND WALSH, LIMITED,

YOKOHAMA, SHANGHAI, HONGKONG, AND SINGAPORE.

PREFACE.

In publishing the original edition of this work, the Author expressed a hope that the Japanese method of decorating with flowers might be found adapted to adorn our Western homes. He felt assured that the æsthetic rules governing this Floral Art, though novel to us in their application to flower decorations, would, nevertheless, appeal to European taste as true art principles derived from a close study of natural laws, and not merely as the outcome of a quaint and capricious fancy from the Far East. A reviewer, in noticing the first edition, contended that the art theories expounded could not claim novelty or originality since they were universally known to all designers. Now, it was just the appreciation of this fact that led the present writer, an architect by profession, to devote his attention to the subject treated. In our buildings, our furniture, our decoration, and our industrial arts, the tendency of the age is to discard false and meaningless designs, and to follow the true principles laid down by Ruskin and other contemporary art teachers. If floral cuttings are to be used in adorning a room, why should we not apply similar art principles to their employment? Why should flowers alone be used in disorderly confusion, whilst order and method rule in other things? Working with stone, brick, timber, metal, paint, or plaster, we aim at expressing the true qualities of the material, both in construction and ornamentation. Yet when our material is the freshest and loveliest that the earth yields we are content to use it in a disorderly and expressionless manner. Cut flowers, as we arrange them, retain no suggestion of their natural growth or of the landscape to which they belong. With all our passion for floriculture and for masses of rich bloom, we have in Europe never possessed a Floral Art, properly so-called, an art of designing with plant and tree cuttings in such a way as to conventionally copy Nature in her disposal and treatment of floral growth. Mrs. C. W. Earle, in her charming work "Pot pourri from a Surrey Garden," has expressed her admiring appreciation of the Japanese mode of arranging flowers for embellishing rooms, and has given several examples showing how such a method may be applied to English flowers. There are many indications that the study of this Floral Art is growing abroad, and that it will continue to influence our use of cuttings for chamber decoration. The writer does not suppose that the style of flower arrangements followed by the Japanese will ever be adopted complete, with all the quaint traditions and subtle ethics that surround it in the land of its origin, but an explanation of these details is desirable in order to show the *rationale* which evolved the Art in the hands of this cultured people, and to indicate the lines upon which a suitable European adaptation may be based.

The above considerations encourage the Author in issuing a second and revised edition of his original work. Additional outline plates and figures in the text have been introduced, as well as new coloured prints from designs expressly made by Ogata Gekko, one of the best living artists of the Popular School. The text has been augmented to some extent, partly in the body of the work, and partly in an appendix. The somewhat lengthy title of the first edition has been dispensed with, and the abbreviated one "Floral Art of Japan" adopted.

JOSIAH CONDER.

Tokio, May, 1899.

CONTENTS.

DESCRIPTION OF ILLUSTRATIONS.

COLOURED PLATES.

OUTLINE PLATES.

ILLUSTRATIONS PRINTED WITH THE TEXT

THE FLOWERS OF JAPAN.

INTRODUCTION.

"Flowers seem intended for the solace of ordinary humanity: children love them; quiet, tender, cultivated, ordinary people love them as they grow; luxurious and disorderly people rejoice in them gathered."—John Ruskin.

ONE of the general impressions which exist abroad with regard to Japan, is that of a land unrivaled in the abundance and beauty of its flowers,—a belief that nature has lavished her floral gifts with special favour upon these sunny islands of the Far East. And in a great measure is this popular fancy justified, considering the imposing display of flowering trees and shrubs to be seen near the principal cities at certain seasons. In the sense, however, of profusion in wild floral plants, it must be admitted that certain Western countries possess attractions which Japanese scenery can scarcely boast. The traveller, whose rural wanderings nearer home have made him familiar with furze and heather-clad moors, flower-sprinkled meadows, or hills and forests girdled and carpeted with blossoming plants, will miss in Japan some of these charming adornments of natural landscape. There is one short season in the year,—when the rice is young and green, and the bloom of the honey-scented rape plant spreads broad stretches of yellow on the plains,—that recalls to the mind the soft flowering verdure of other lands; but the rice culture, with its endless irrigated terraces, and the high bamboo-grass, which covers every uncultivated hill and glen, deprives the scenery of all but a brief and passing suggestion of the colouring of Western meadows and uplands. Tiger lilies, fratillaries, bluebells, and numerous other wild flowers grow on the plains and downs, buried in the high coarse bamboo-grass. The comparative scarcity of groups of wild flowering plants, as a colour feature to the landscape, is, however, to some extent made up for by the blossoming trees, displaying in turn soft and vivid masses of colour amid the verdant foliage of the hill-sides. The wild camellia, azalia, magnolia, plum, peach, and cherry are the most important of these flowering trees, the varied flaming tints of the azalia contributing chiefly to the brightness of the scenery. The blossom of the wild cherry tree, which abounds in

Yezo, adds to the wooded landscape of this Northern island an effect as of soft clouds of a pale pearly tint, which the poets liken to mist upon the mountains.

The flower charm as it exists in Japan is not, as elsewhere, purely one of rural or pastoral associations. It is closely and inseparably connected with the national customs and the national art. The artistic character of the Japanese people is most strikingly displayed in their methods of interpreting the simpler of natural beauties. That extravagant taste which demands for its satisfaction the grand, rare, or novel, and is unmoved by the modest attractions of more familiar surroundings, is little shared by the masses of this people, who bestow their chief sympathies on the humbler aspects of nature. Though the more majestic scenery of the country has served as a constant theme for the painters and poets of the Court and nobility, the *popular* art has always been an expression of the daily life of the people, and of those simple, natural surroundings, customs, and familiar beliefs which contribute to its enjoyment. The recurring months of the year, and the various duties, pleasures, and poetic fancies which custom has associated with them, form the inexhaustible source from which artists of all kinds have looked for inspiration. The common flowers of the seasons have been given a prominent place in the fête-day calendar. Almost every month is known by its special blossoms, and the important cities have groves and gardens devoted to their public display. The festivals of the numerous Shinto deities are celebrated monthly in the different towns by street fairs at which the choicest specimens of flowering trees and plants are exposed for sale. The secret, then, of Japan's floral fame and floral enchantment lies rather in the care that her people bestow upon Nature's simpler gifts than in any transcendent wealth of production. Treasured chiefly as heralds of the seasons, and as inseparable from the favourite pursuits and pastimes of out-door life, Japanese flowers are by no means esteemed in proportion to their scarcity or difficulty of culture. The isolated merit of rarity, so much sought after in the West, has here little or no attraction. The native florists are not deficient in floricultural skill, and produce in certain blossoms forms of considerable artificial beauty, but the popular taste shows a partiality for the more ordinary and familiar flowers, endeared by custom and tradition.

Flower viewing excursions, together with such pastimes as *Shell gathering*, *Mushroom picking*, and *Moon viewing*, form the favourite occupations of the holiday seeker throughout the year. By a pretty fancy, even the snow-clad landscape is regarded as Winter's floral display, and *Snow viewing* is included as one of the *flower* festivals of the year. The Chinese calendar, used formerly by the Japanese, fitted in admirably with the poetical succession of flowers. *Haru*, the Japanese Spring, opened with the New Year, which commenced about February, and was heralded by the appearance of the Plum blossoms.

SPRING FLOWERS.

SPRING FLOWERS.

PLUM BLOSSOMS.

ENRICHING the bare landscape with its bloom, and filling the air with its fragrance at a time when the snow of winter has hardly passed away, the blossoming plum tree has come to be regarded with especial fondness by the Japanese. Combined with the evergreen pine and bamboo, it forms a floral triad, called the *Sho-chiku-bai*, supposed to be expressive of enduring happiness, and used as a decorative symbol on many felicitous occasions. The plum blossom being the earliest to bloom in the year, is often referred to as "the eldest brother of the hundred flowers." Quick in seizing the peculiar features which distinguish one growth from another, to the extent almost of a tendency to caricature them, the Japanese have been chiefly attracted by the rugged and angular character of the plum tree, its stiff, straight shoots, and sparse, studded arrangement of buds and blossoms. Thus, a fancy has arisen for the oldest trees, which with their gnarled trunks and tender shoots exhibit these characteristics to perfection. In them is shown the striking contrast of bent, crabbed age, and fresh, vigorous youth. As if to render this ideal more complete, it is held that the plum tree should be seen in bud rather than in full blossom.

The gardeners of the country, so clever in the training of miniature trees, find in the plum a choice object for their skill, imitating on a small scale this favourite character of budding youth grafted on to twisted and contorted age. These tiny trees, trained in a variety of shapes,—bent, curved, and even spiral,—with their vertical or drooping graftings of different coloured blossom-sprays, fresh, fragrant, and long lasting, form one of the most charming of room decorations during the first months of the year.

Poets and artists love to compare this flowering tree with its later rival, the cherry. With the latter, they say, the blossom absorbs all interest; whereas, in the case of the former, attention is drawn more to the shape of trunk and branches: the cherry blossom, it is not denied, is the prettier and gayer of the two; but the plum blossom is more chaste and modest in appearance, possessing besides, its sweet odour. Enthusiastic admirers of the cherry blossom, however, go so far as to fancy they detect a delicious odour in this scentless flower. The fragrance of the plum blossom is constantly referred to in the short stanzas with which the poetry of the country abounds. The following free translation may be given as an example of one of such verses :—

"In Spring time, on a cloudless night,
When moon-beams throw their silver pall
O'er wooded landscape, veiling all
In one soft cloud of misty white,
T'were vain, almost, to hope to trace
The plum trees in their lovely bloom
Of argent, t'is their sweet perfume
Alone which leads me to their place."

The custom of planting plum trees in groves and avenues to form pleasure resorts during blossom time, seems to be of comparatively modern date; and some of the most famous plum groves were originally orchards, planted for the sake of the fruit. It is said that in China, from whence Japan borrowed many of her customs and cults, this tree was first esteemed for its fruit alone, and in later and more æsthetic times it became honoured for its pure blossom and sweet scent. In the earliest Japanese annals we read of a single plum tree being regularly planted in front of the South pavilion of the Palace at Nara, and of its being replaced by a cherry tree in later times, when the latter had in its turn become the favourite of the Court. In connection with this Imperial custom, a pretty story is told explaining the origin of the name *O-shuku-bai*, or *Nightingale-dwelling-plum-tree*, applied, even to the present day, to a favourite species of delicious odour having pink double blossom. Some time in the tenth century the Imperial plum tree withered, and, as it was necessary to replace it, search was made for a specimen worthy of so high an honour. Such a tree was found in the garden of the talented daughter of a famous poet, named Kino Tsurayuki, and was demanded by the officials of the Court. Not daring to resist the Imperial command, but full of grief at parting with her favourite plum tree, the young poetess secretly attached to its trunk a strip of paper upon which she wrote the following verse :—

"Claimed for our Sovereign's use,
Blossoms I've loved so long,
Can I in duty fail?
But for the nightingale
Seeking her home of song,
How shall I find excuse?"—Brinkley.

PLATE I.

PLUM BLOSSOMS AT SUGITA.

This caught the eye of the Emperor, who, touched by the plaintive sentiment expressed, enquired from whose garden the tree was taken, and ordered it to be returned. The season of the plum blossom is made musical with the liquid note of the so-called Japanese nightingale (*Cettria cantans*), which for this reason is inseparably associated with the plum tree in the different decorative arts. Such dual combinations, taken from the animal and vegetable kingdom, form favourite motives for designs. Bamboo leaves with Sparrows, Pea-fowl with Peonies, Tigers with Bamboos, Storks with Pine trees, Wild boar with Autumn grasses, and Deer with Maples, may be mentioned as some of these popular combinations.

In later times plum trees were planted in large numbers at rural spots near to the Imperial capitals, forming pleasure resorts for the ladies of the Court. Along the banks of the river Kizu, at a place called Tsuki-ga-se, in the province of Yamato, fine trees of pink and white blossom extend upwards of two miles, diffusing their delicious scent around: they are what remains of quite a forest of plum trees said to have stretched for miles around. The more modern towns have also their favourite plum orchards, visited by crowds of sigh-tseers at blossom time, in February. Sugita, a village not far from Yokohama, possesses one of the most famous; having over a thousand trees, many of which are from eighty to a hundred years of age, and which supply in the Summer most of the fruit consumed in the Eastern Capital, Tokio. It is popularly known and frequented on account of its blossoms in the early Spring, and boasts six special varieties of tree, distinguished by different fancy names having reference to the character of the flower; the principal of which are those of pink and of so-called green blossom,—for the white kind has a faint tinge of emerald.

In all, there are said to be sixty different species existing in Japan. The single blossom of white or greenish-white colour and of small size is held most in esteem. All the white kinds are scented, but of the red some possess no perfume. There is an early plum of red and double blossom which blooms before the Winter solstice, and is of handsome appearance, but it has little or no scent. The Japanese include several species of the *Jasminum* in the same category as plum trees.

Every visitor to Japan has heard of the *Gwa-rio-bai*, or Recumbent-dragon-plum-trees at Kameido, a famous spot in the North of Tokio. At this place there existed, up to fifty years ago, a rare and curious plum tree of great age and contorted shape. Its branches had bent, ploughing the soil, and forming new roots in fourteen places, and it straggled over an extensive area. Owing to its writhing and suggestive shape, it received the name of the Recumbent Dragon, and, yearly clad with fresh shoots and white blossom of fine perfume, attracted large crowds of visitors. From this famous tree, fruit is said to have been yearly presented to the Shogun. Succumbing at last to extreme age, it has been replaced by a number of less imposing specimens, selected on account of their more or less bent and crawling shapes. The present group of plum trees, inheriting the name though but little of the character of Recumbent Dragons, makes a fine show of blossoms in February, and keeps up the popularity of the resort.

Komurai and Kinegawa, near Kameido, also have blossom-groves which are much frequented.

Another noted spot, is Komukai, near Kawasaki, not far from Tokio, which is historically famous as having been often visited by the Shogun, and possesses trees over two hundred years of age.

At Shinjuku, another suburb of Tokio, is a fine grove of plum trees, sometimes called the Silver-world (*Gin-sekai*), a term applied to the snow-clad landscape, and having special reference in this instance to the silver whiteness of these blossoms.

The favourite trees of single blossom are eight in number—the Hitoye-ume, Shibori, Hosoka, Nishikin, Kotenbai, Shidare, Suzuri-shidare, and Tokiwa-shidare; and among the most prized of double blossom are the Shidare-yaye, Suzuri-yaye, Okina-ume, Beni-suzume, Yunera, and Hana-gomori. Other trees such as the Mangetsu, Kaoru-ume, Momochidori, Tani-no-yuki, and Miyakodori are known as the best fruit-bearing kinds of plum tree. The illustration, Plate I, represents the plum grove at Sugita, about five miles from Yokohama, which can be approached from the sea beach.

PEACH BLOSSOMS.

QUICKLY after the plum follows the peach blossom which, though by no means sharing the traditional esteem and admiration bestowed upon the former, excels it in size, richness, and colouring. These blossoms are of numerous tints,—white, and different shades of pink, and a deep crimson remarkably rich in tone. The peach blossom in mass, as it appears in groves and orchards, contributes far more to the beauty of the

PLATE II.

PEACH BLOSSOMS: SOKA NO MOMOYAMA, SENJU.

Spring landscape than its more honoured but severer brother the plum blossom, though it has not, however, had the good fortune to be patronized by ancient Emperors or extolled by distinguished poets. Lacking these essentials to floral greatness, and coming as it does between the plum, of classical fame and predilection, and the flashy cherry of patriotic boast, it has been comparatively neglected by the artist and relegated to secondary rank as a decorative motive and material. The orchards of peach trees in blossom are, however, much frequented by the common people who find enjoyment at any spot where bloom and colour are to be seen.

Plate II. illustrates trees in a grove called Soka-no-Momoyama, at Senju, a suburb of Tokio.

CHERRY BLOSSOMS.

THE third month of the old Japanese Spring, corresponding with the present April, is the month of the cherry blossom, the king of flowers in Japan. This flower is remarkable for its softness and exuberance, as contrasted with the severe simplicity of the plum blossom. The latter blooms fresh, vigorous, and leafless, in the bare and often snow-clad landscape; the former, with its florid richness, enchanced in some cases by young reddish leaves, is especially fitted to assert itself amid the greenery of budding spring-time. But the splendour of the cherry's bloom is transitory in comparison with the more lasting qualities of the plum, which retains its beauty for a full month. The cherry flowers must be viewed during the few short days of their prime, and should these days be stormy, the full glory of the sight is lost for a year. The most enthusiastic partizans of the cherry blossom assert that it is all the more precious on account of its transient character. "Among men, the samurai, among flowers, the sakura," is a familiar saying, which well expresses the patriotic pride with which this gay flashy flower is regarded in Japan. The sentiment is also conveyed in the following lines by one of the national poets:—

"Shikishima no
Yamato gokoro wo
Hito towaba
Asahi ni niwou
Yamazakura bana."

"Should you ask me what is the true patriotic spirit, I answer—it is the scent of the mountain cherry tree in the morning air."

The wild cherry seems to have existed in Japan from time immemorial, and still abounds in the woods of the Northern island, Yezo, where the Aino aborigines apply its bark to many purposes. In ancient times, however, the plum tree, of Chinese importation, seems to have absorbed the attention of the Court and people, and it was at a later date that the cherry, the flower of the country, appears to have found its place in their affections. Though early records refer frequently to the plum, there is no mention of the cherry earlier than the time of Richiu, an Emperor of the fifth century. This Monarch was disporting himself with his courtiers in a pleasure boat, on a lake of the Royal park, when some petals from the wild cherry trees of the adjoining hills fluttered into the wine cup from which he was drinking. This circumstance is said to have drawn His Majesty's notice to the beauty of this neglected blossom, and from this time arose the custom of wine drinking at the time of cherry viewing. To the present day there is a popular saying,—"Without wine who can properly enjoy the sight of the cherry blossom?" It was reserved for a later Emperor, in the eighth century, to give it that importance as a national flower which it has ever since retained. Whilst on a hunting expedition on Mount Mikasa, in the province of Yamato, the Emperor Shomu, attracted by the beauty of the double cherry blossoms, composed the following short verse, which he sent, with a branch of the flowers, to his favourite Consort, Komio Kogo:—

"This gathered cherry branch can scarce convey
A fancy of the blossom-laden tree,
Blooming in sunlight; could I show it Thee,
Thoughts of its beauty would drive sleep away!"

To satisfy the curiosity of the ladies of his Court, the Emperor ordered cherry trees to be planted near the Palace at Nara, and afterwards the custom was continued at each succeeding capital.

Yamato, the province in which were situated several of these ancient capitals, is the most noted for its cherry groves, and at a spot called Yoshino a thousand trees line the path and cover the hill side. It has been a favourite fancy to compare the appearance of these trees in blossom to mists or snow upon the hills as in the verse,—"The cherry blossoms on Mount Yoshino deceive me into thinking they are snow!"

PLATE III.

CHERRY BLOSSOMS AT MUKOJIMA.

Imperial garden parties to view the national flower date back as early as the days of the Emperor Saga, in the ninth century. These ancient court receptions were attended by the notable *literati* whose amusement was to compose odes on the flowers. In the thirteenth century, the Emperor Kameyama caused a number of trees from Yoshino to be planted at Arashiyama, a beautiful hilly spot on the banks of the rapid river Oi, in the neighbourhood of Kioto. Here he built a summer pavilion, and, in spring and autumn, Court after Court visited the lovely spot, which was rendered further famous in a verse composed by one of the Imperial line:—"Not second to Yoshino, is Arashiyama, where the white spray of the torrent sprinkles the cherry blossoms." This spot no longer possesses its Imperial pavilion, but remains a favourite resort for sight-seers from the Western capital, in the months of the cherry and the reddening maple. Numerous tea houses and booths, on the banks of the rapids, give a fine view of the wooded hills opposite, amidst the Spring greenery of which may be seen the pearly white clouds of the cherry blossoms. Here the blossom-clad branches form a part of the distant landscape, as originally beheld in their natural wildness, when they first attracted the admiration of the earlier Emperors, and before their more gorgeous successors, of double-flower, became arranged in artificial groves and avenues.

In and near to the modern capital, Tokio, are several spots renowned for their show of cherry trees, originally brought from Yoshino, and from the banks of the Sakura river in the province of Hitachi. It is said that cherry viewing first became a popular amusement in Yedo towards the latter half of the seventeenth century. From that time all classes of the people, from the two-sworded samurai to the small tradesmen and menials, participated in the enjoyment. The green sward beneath the cherry trees was crowded with merry picnic parties of all classes, screened off with low coloured curtains. One favourite resort, called Asukayama, at Oji, is often spoken of as the *new* Yoshino. It is a high, grassy bluff, overlooking an extensive plain on one side, and sloping down to the road of the Oji village on the other. The eminence forms a park of cherry trees extending down the precipitous sides of the bank, so that the pedestrian sees the pinky white blossoms against the blue sky above him, and below him the pearly gray of the blossoms in the shadow of the cliff.

Koganei, some half day's ride from Tokio, is perhaps the most attractive spot for seeing the double cherry in full bloom. Here a fine avenue of these flowering trees extends upwards of two and a half miles along the aqueduct which conveys the water of the river Tama to Tokio. It is said that they were first planted immediately after the completion of the aqueduct, by command of the Shogun Yoshimune, in the beginning of the eighteenth century, with the idea that cherry trees had the virtue of keeping off impurities from water. For this purpose ten thousand trees were brought from Yoshino and from the banks of the river Sakura; but the number now remaining has dwindled to only a few hundred.

In the old temple grove, now a public park, at Uyeno, there are a number of fine trees of the single early-blossoming kind, called by the Japanese *Higan-zakura*, among which are some magnificent specimens of the weeping cherry. This latter species has pendant branches, drooping like the willow, and bears single white flowers, but no fruit; and in this respect it is an exception to the general rule, that the trees of single blossom bear fruit whilst those of double blossom are fruitless. The fruit of the Japanese cherry tree is, however, at its best, insipid and worthless. These trees at Uyeno are said to have been planted by one of the Tokugawa Regents in imitation of the hills at Yoshino; they are all of majestic size, and present a gorgeous sight in April, with their pale pink blossoms seen partly against the blue sky, and partly against the rich foliage of the pines and cedars which surround the golden shrines and cenotaphs of the Shoguns. The single-blossom trees at Goten-yama, a park in the suburb of Shinagawa, form a beautiful sight early in April.

The most popular resort in Tokio is the cherry avenue at Mukojima, extending for more than a mile along the banks of the river Sumida. Here the trees lack the grandeur and natural beauty of those at Uyeno, and have no surrounding foliage to set them off; but they are mostly of double blossom, and bending with their weight of flowers, —looking almost artificial in their luxurious fullness,—present a most imposing sight. This spot is frequented by the gayest holiday makers. Wine drinking is considered essential to a proper enjoyment of the scene, and crowds of pedestrians, bearing their gourds of rice-wine, make such resorts merry and boisterous with their carousals. Other visitors, of a richer class, indulge in the prospect of the blossom-laden banks from roofed pleasure boats, accompanied often by a gay gathering of singing and dancing girls.

The season of this flower is one of high winds, and the soft petals of the full blown blossoms fall like snow flakes covering the pathways beneath. This simple fact is not without its attraction to the Japanese, who make much of the falling cherry petal in their poetry and other arts.

"No man so callous but he heaves a sigh
When o'er his head the wither'd cherry flowers
Come flutt'ring down. Who knows? the Spring's soft show'rs
May be but tears shed by the sorrowing sky."—CHAMBERLAIN.

The cherry trees in blossom, seen at night by the pale light of the moon, form also another great attraction. *Yozakura* or Night Cherry Flowers being included as one of the sights of the year. The river banks at Mukojima and even the formal avenues of the Yoshiwara at Asakusa are crowded after sunset, in the cherry season. Cherry viewing at Mukojima on the banks of the Sumida river is illustrated in Plate III., which represents excursionists ascending the banks from a pleasure boat.

SUMMER FLOWERS.

WISTARIA BLOSSOMS.

ONE of the earliest flowers of the Japanese Summer which attracts the pleasure seeker, is the wistaria, blooming in May, soon after the cherry blossom has fallen. This stalwart flowering creeper is reared upon large trellises, arranged to cover long walks, bridges, or arbours, in pleasure grounds and gardens. A favourite position is one sheltering an open gallery, which overhangs a lake or stream. In the precincts of the popular temple at Kameido, in Tokio, close to the famous Recumbent Dragon plum trees there are wistarias of magnificent size, bearing blossoms which hang in rich purple trails from two to three feet in length. Wide rustic galleries, in connection with matted garden sheds and galleries, extend over an artificial lake stocked with gigantic gold fish, and the wistaria trellises form an extended covering overhead. A belief exists that this flower attains great size and beauty if its roots are nourished with the rice-wine of the country, and there is, at Kameido, a tree producing specially fine blossoms, at the base of which visitors are accustomed to empty their wine cups. Fine specimens exist in various parts of Japan, bearing clusters over three feet in length, among which may be mentioned one at Noda, in the province of Settsu, called the *Chitose*, or tree of a thousand years.

The wistaria of purple blossom is most common and at the same time most esteemed, ranking higher than the white kind, which is regarded as abnormal. This is an exception to the prevailing custom, which places white before other colours in blossoms of the same species, and especially proscribes purple flowers as associated with mourning, and, therefore, unfit for felicitous occasions. In various designs the pheasant is shown in combination with the purple wistaria.

The view of the wistaria in flower as enjoyed in olden times by ladies of rank is shown in Plate IV.

SUMMER FLOWERS.

WISTARIA BLOSSOMS IN A NOBLE'S GARDEN.

AZALIAS IN TEA HOUSE GARDEN, NEZU

AZALIAS.

THE azalias commence to flower about the same time as the wistarias, and display a variety of colours of most brilliant hue—numerous shades of scarlet, crimson, orange, cream, white, and magenta—unequalled by any other blossom. The hardy azalia shrubs are abundantly planted on the sides of artificial hillocks and on the slopes of terraces, and a Japanese landscape garden, usually remarkable for its wealth of evergreens and predominating verdure, never looks gayer than when these bushes are in flower. There are several public gardens at Okubo, a village near Shinjuku, in Tokio, which are planted entirely with azalia bushes of great size and remarkable age. These plantations date back to the time of the Tokugawa Regents, by whom they were frequented, and they are still visited every summer by numbers of sight-seers. One azalia tree at Okubo has a stem as thick as a man's leg and is said to produce eight thousand blossoms at a time. Other places where the azalias may be seen to advantage are,—Uyeno Koyen, Uyeno Okeiyen, Asakusa Koyen, Shiba Koyen, Susaki Benten-no-sha-nai, Mukojima Mokubo-ji, Azabu Shokayen, Honjo Uyebun, Meguro Daikokuya, and Horikiri Musashiya,—all parks or gardens in the vicinity of Tokio. Plate V. illustrates the azalias in the grounds of the Gongen shrine at Nedzu, a suburb of Tokio. In this enclosure is a famous suite of chambers used for the Tea Ceremonial and much patronized by the public.

IRISES.

IN June, the popular flower is the iris or flag, which is cultivated in large marshy flats near to rivers or lakes. In many gardens, watered by a stream, a loop or bend in the water-course is spread out into a marshy expanse, planted with irises, and crossed by fancy plank bridges of zigzag shape. There are in Japan four distinct species of iris, known by different native names, but the kind most seen is the *Acorus calamus*, or sweet flag, which the Japanese call *Hana-shobu*. In the case of displays of iris flowers, a mass of varied colour is delighted in; the purple, white, and variegated blossoms being grown together, indiscriminately, and with little or no attempt at pattern or design. The most noted place for shows of this flower is Yatsuhashi, in the province of Mikawa, the scenery of which locality serves as the model for the iris beds of gardens; but the popular resort nearest Tokio is a spot called Horikiri, close to the river Sumida, to which place it is the fashion to make excursions in pleasure boats early in June. Here the beds which contain the flowers in every variety of colour, are surrounded by elevated grassy banks, dotted with summer-houses, from which visitors can look down upon the richly variegated carpet below. Narrow wooden bridges give further picturesqueness to the scene, crowded in the season with a brilliant throng of visitors, whose pretty costumes almost vie in gaiety of colour with the flowers.

The iris, as a water plant, is associated in art with the kingfisher, water rail, mandarin duck, and other water birds.

Plate VI. shows the iris beds as they may be seen at many places on the outskirts of the city.

PEONIES AND LOTUS FLOWERS.

AMONGST summer flowers must be noticed the peony and lotus, which, though hardly sufficiently democratic to rank among the most popular flowers, yet play an important part in the art of the country. The peony is a delicate plant and is cultivated in long sheltered beds, generally forming the *parterre* to some adjoining chamber, from which its magnificent blossoms can be viewed. In the grounds of the wealthy it is subjected to scrupulous care and nursing, in order to produce flowers of enormous size and fullness, often so large and heavy as to need artificial support. It is regarded as the flower-queen of China, and is essentially the favourite of the upper classes in Japan. The peony was first imported into this country in the eighth century, and was then chiefly cultivated in the provinces of Yamato and Yamashiro. Even now, the finest specimens in Tokio are brought from the neighbourhood of the old capital, Nara. The largest blossoms measure as much as nine inches across. The peony is sometimes called the flower-of-prosperity; another fancy name by which it is know is the plant-of-twenty-days, because it is said to preserve its beauty and freshness for that period of time. Of the large tree-peony, called *botan*, there are ninety distinct kinds, and of the small plant-peony, having

PLATE VI.

IRIS BEDS, NEAR TOKIO.

PEONIES AT YOTSUME, HONJO.

single blossoms and called *shakuyaku*, there are said to exist five hundred varieties. The *botan* may be inspected at numerous public gardens in Tokio such as the Hanjitsuyen, Unsanyen, Gochikuyen and Sendaiyen at Komagome; the Uyebun, Yokayen, and Teigoyen at Honjo; Daikokuya, Meguro; Taikoyen, Shiba; and Senkayen and Shokayen in Azabu. The *shakuyaku* is also shown in the Uyebun, Honjo; the Yoshinoyen at Komme-hikifune-dori and the Teigoyen at Minami futaba-cho Honjo. Among colours, the red and white are most valued, purple and yellow specimens, though rare, being less prized. This exuberant flower, with its large curling petals, is a favourite subject for design and decoration. Its companions in art are the peacock, the golden pheasant, and the *shishi*, a kind of conventional lion derived from Chinese designs; in such company it forms the constant decoration of temple and palace walls. A show of peonies in the screened and sheltered beds in which they are grown is represented in Plate VII, taken from the garden of the Uyehan, Yotsume, Honjo.

The lotus is closely connected with the Buddhist religion, and is, therefore, associated in the minds of the people with spirit-land. The lakes of temple grounds, especially those dedicated to the water goddess Benten, are frequently planted with lotuses. The lake Shinobazu at Uyeno has a fine display. The fine wide moats of the Tokio Castle abound in these water plants, which impart to them considerable beauty in the summer season. Wherever undisturbed pools and channels of muddy water exist, the lotus is to be found, and even the ditches beside the railway connecting Tokio with the port of Yokohama are rendered gay in the summer by the lotus flowers in bloom. As the peony is said to be the national flower of China, so the lotus is regarded as the national flower of India, the source and centre of Buddhism. It is therefore considered out of place as a decoration for occasions of festivity and rejoicing, but it is constantly used for obsequies and other sacred ceremonies. The lotus serves as a suitable theme for religious contemplation, and is therefore the favourite flower of monastic and temple retreats: the best displays are to be seen in the lakes of the old temple groves of Kioto and other cities. Growing out of the muddiest and most stagnant water, its leaves and flowers are always fresh and clean; although it is particularly sensitive, and quickly withers if brought in contact with any of the fertilizers by which other plants are nourished. This purity which the lotus maintains amid surrounding filth is mentioned as one reason for associating it with a religious life. A well-known book of Buddhist precepts contains this text:—"If thou be born in the poor man's hovel, but hast wisdom, then art thou like the lotus flower growing out of the mud!"

The white lotus flower has a powerful and sweet perfume, but the red kind, though more handsome, produces but little scent. There is a species called *Gold-thread-*

lotus, its red blossoms being marked with yellow lines; and a very handsome flower of deep crimson colour may also be found. The Indian lotus, which is sometimes to be seen in Japan, has a large double flower, of red colour, which never closes day or night, but falls off after five or six days. The blossoms of the ordinary kinds close after midday. The leaves of the lotus are almost as much esteemed as the flowers, their broad curling surfaces of deep green and emerald presenting a beautiful effect in the lotus ponds, and forming a favourite subject for the painter's brush. In designs, the mandarin duck and other water fowl are represented with the lotus.

AUTUMN FLOWERS.

AUTUMN FLOWERS.

CHRYSANTHEMUMS.

THE chrysanthemum is the principal flower of Autumn, and the triumph of Japanese floricultural skill. Specimens of remarkable variety in form and colour of blossom are produced in the gardens of the nobility. The flower of the chrysanthemum, in its most handsome form, loses its disc-like character, and presents a combination of long oval petals, partly extended, and partly curling inwards, exhibiting in contrast the different tints of face and back; whilst, in its most eccentric and artificial shape, it assumes the character of a confused mop of tangled thread-like petals, more curious than beautiful. The florists aim at producing an extraordinary quantity of blossoms upon one stem, reaching often to the number of several hundreds. The varieties of the Japanese *kiku* are numerous, including not only those species classified as chrysanthemums by European botanists, but many kinds of Pyrethrum, Aster, and other genera.

It seems that the wild chrysanthemum, of small flower, has always been indigenous to Japan, and held in considerable repute from every early times for medicinal purposes; in which connection early records state that large quantities of the yellow kind were yearly sent to the Imperial Court from the Southern provinces. The large cultivated flower, however, is said to have been imported from Corea or China and first planted in Japan at Hakata in the province of Chikuzen. At this time five colours were known, described as blue, yellow, red, white, and black, the term *black* probably referring to a dark purple colour. Originally these plants were reproduced by means of slips and cuttings, but now the seed is employed, which is said to give greater variety of blossom.

The chrysanthemum is sometimes spoken of by foreign authors as the national flower of Japan, a rank properly belonging to the cherry blossom; and this mis-conception is probably owing to the former being used as one of the crests of the Imperial House The flower has always been much honoured by the Court, and as early as the time of the

Emperor Heizei, in the ninth century, garden parties were held in the Palace for the purpose of celebrating its blossoming time: just as, at the present day, a yearly chrysanthemum show takes place in the Imperial grounds. These ancient celebrations seem to have partaken of a truly pastoral character, the courtiers wearing the plucked blossoms in their hair, drinking wine, and composing verses upon the beauties of the flowers. The modern chrysanthemum displays in the Palace gardens are more like our own flower-shows in the social conventionality of their arrangements; but the numerous variety, of every imaginable colour and profusion of shape, arranged in long open rustic sheds, forms a brilliant and imposing scene hardly rivalled by any flower-show in the world.

At a recent display in the Imperial grounds at Akasaka there were a hundred and sixty varieties of blossom exhibited, each bearing a fancy name of its own. Some of these names are so poetically suggestive of the form or colour of the flowers that the enumeration of a few of them will not be out of place:—

Chi-kiu-gi .	Terrestrial Globe—a large globular yellow flower.
Gin-sekai .	Silver World—a flower of pure white colour.
Usu-gasumi . .	Thin Mist—a white flower.
Tsuki-no-tomo . .	Companions of the Moon—a white flower.
Yu-hi-kage	Shadows of the Evening Sun—a flower of dull red colour.
Tama-sudare .	The Screen of Gems—a flower of orange red colour.
Hatsu-yuki .	The First Snow—a pure white blossom.
Hana-gatami . .	The Basket of Flowers—a rich red flower.
Kagari-bi .	Beacon Light—a red flower.
Asa-hi-no-nami .	Waves in the Morning Sun—a reddish flower.
Ake-no-sora	Sky at Dawn—a flower of cherry-blossom colour (pale pink).
Shigarami . . .	Garden Fence—a flower the colour of the wistaria blossom (lavender colour.)
Asa-ne-gami . .	Dishevelled Hair (lit. in morning sleep)—a flower of tangled petals.
Hoshi-dzuki-yo . .	Starlight Night—a white flower.
Hoshi-no-hikari . .	Star's Brightness—a pale bluish flower.
Kimi-no-megumi	Blessings of Majesty—a pale pink flower.
Yuki-no-ashita . .	Snowy Morning—a flower of pale pearly pink colour.
Tsuki-no-kasa . .	Moon's Halo—a flower of orange red colour.
Ogon-no-nishiki . . .	Golden Brocade—a flower of golden yellow colour.

CHRYSANTHEMUMS AT SOMEI.

Shimo-no-ha	Leaves in Frost—a white flower.
Ogon-no-tsuyu	Golden Dew—a bright yellow flower.
Mangetsu . .	Full Moon—a white flower.
Gek-ka-no-nami	Moonlit Waves—a flower of yellowish white colour.
Haku-rio . .	White Dragon—a white flower.
Tsuyu-no-shita-zome	Dye of the Dew—a flower of pale yellow colour.

There are said to be in Japan two hundred and sixty-nine colour varieties of the chrysanthemum, of which sixty-three are yellow, eighty-seven white, thirty-two purple, thirty red, thirty-one pale pink, twelve russet, and fourteen of mixed colours. A fancy prevails that in this flower the same tint is never exactly reproduced, and that it thus suggests the endless variety of the human countenance. Blooming longer than most flowers, the chrysanthemum has come to be associated with longevity. In the province of Kai, a hill, called the Chrysanthemum Mount, overhangs a river of clear water, into which the petals fall, and a belief exists that long life is assured by drinking the water of this stream. A favourite motive of decoration, which may be seen in numerous conventional designs, is the chrysanthemum blossom floating in running water. A custom also survives of placing small blossoms or petals in the cup during the wine-drinking which takes place on the festival of the ninth day of the ninth month.

The ordinary varieties of chrysanthemum are to be seen in great abundance in the street fairs during the Autumn months. Dango-zaka, in Tokio, is a favourite popular resort during the chrysanthemum season, but here the flowers, mostly of the smaller kind, are used artificially, modelled into groups of figures and animals representing historical subjects, scenes from popular theatrical performances, and even the battle scenes of the China-Japan war.

The chrysanthemum is associated with the crane, the royal bird of Japan. Plate VIII. illustrates a display of chrysanthemums in a gardener's grounds at Somei.

THE SEVEN PLANTS OF AUTUMN.

THE paucity of important flowering trees and plants in Autumn, has led the Japanese to make much of certain simple plants, comparatively insignificant in themselves, but gathering importance and interest in combination. As has already been pointed out, almost every month of the year is associated with a special blossom, and the calendar would therefore not be complete without a reference to these flowers of the late Autumn. These seven plants are: the lespedeza, the morning glory, the *Eularia japonica*, the *Valeriana villosa*, the *Valeriana officinalis*, the *Pueraria thunbergiana*, and the carnation. Perhaps the favourite of these is the lespedeza, or bush clover, of which there are several kinds, some having pink, some white, and others yellow flowers. Growing wild on grassy moors, it is associated with wild horses, deer, and the wild boar, together with which it is often depicted in various designs. The deer is specially associated with the Autumn time, and represented also with other Autumn flowers and with the reddening maple. The seven Autumn plants are grown together in the *Hiak-kwa-yen*, or Garden-of-a-hundred-flowers, at Mukojima. The temple grounds of the *Hagi-dera*, near Kameido, are famous for their show of lespedeza flowers. Displays of various kinds of convolvuli or morning glories, trained in pots over skeleton framework representing rustic structures, are also to be seen at the various gardener's nurseries at Iriya, one of the suburbs of Tokio, by those enthusiastic enough to reach the spot before six o'clock on an Autumn morning. Fancy flowers, less than half an inch in size, in clusters, and shaped like a butterfly orchid, and other strange varieties, may be seen. Within the last year or two the culture of convolvuli or morning glories has become very popular, and they may be observed before many dwellings in the city on quaintly designed frames.

Plate IX. illustrates the seven plants of Autumn as grown in the *Hiak-kwa-yen* at Mukojima.

AUTUMN FLOWERS. AT THE HIAKKA-YEN, MUKOJIMA.

MAPLES

A NOTICE of the floral festivals of the year as observed in Japan demands some mention of the maple,—for the reddening leaf of the maple, like the foliage of many other blossomless trees, is regarded as a flower in Japan. The rich tints of the changing leaves of certain deciduous trees, hardly distinguishable from the colouring of blossoming shrubs such as the azalia, form a favourite object of attraction during the Autumn months. The native term *momiji*, which is commonly translated maple, is, strictly speaking, a general name applied to many trees which redden in the Fall. Of the maple itself, there are many varieties, distinguished both by the form of their leaves and the tone of their changing colour. No garden is considered complete without its group of such reddening trees, placed beside some artificial hill towards the West, to receive additional splendour from the setting sun. They are planted on grassy slopes and in valleys, with the object of bringing into one limited prospect the red and golden tints in which the natural scenery of the wooded hills abounds. The grand slopes above the river Oi at Arashi-yama, noted in the Spring time for their show of cherry blossoms, present a fine display of scarlet maple foliage in the Autumn.

At Ko-no-dai, a famous prominence commanding a view of the whole plain of Tokio, there are some magnificent maple trees, noted for their enormous size. A spot called Tatsuta, in the province of Yamato, is renowned for its fine specimens, which line the banks of the river, and are in full glory about the end of October. At Oji, a suburb of Tokio, the slopes of a natural glen between the hills are planted with thick masses of these trees, forming a most romantic spot, where, from the galleries of a rustic arbour, the sight of the foliage in all its burning splendour may be enjoyed. Shinagawa and Meguro, other well known spots in the vicinity of the capital, have also good groups of maples which attract many sight-seers. Picnicking and mushroom gathering are pastimes which accompany the viewing of the maple.

In the poems and pictures of the country the maple is associated with deer.

"How full of sorrow seems the Autumn! when,
In solitary rambles slowly straying,
Amid the russet foliage of the glen,
I listen to the lonely stag's sad baying."

The famous view of maples in the glen called Taki-no-gawa at Oji is given in Plate X. In the distance may be seen the rustic sheds from which visitors enjoy the prospect of the scarlet foliage, whilst in the foreground is shown a girl reading some of the verses attached to the lower branches of the trees.

PLATE X.

MAPLES AT OJI.

ARRANGEMENT OF FLOWERS.

INTRODUCTION.

WITH the Japanese, the arrangement of cut flowers in vessels of various kinds has become a decorative art of considerable refinement, compared to which Western methods of floral composition have the appearance merely of haphazard combinations. The bouquet, wreath, and garland, all depending for their beauty upon the close massing of blossoms and greenery in soft and luxurious confusion, bear no resemblance whatever to the more austere and open compositions of the Japanese which belong entirely to a different phase of art. The fact that many of the most charming flowers of the country are those of trees, the blossom-clad twigs of which it would be difficult to arrange in closed and rounded masses, may in some manner explain the open lineal character given to floral designs; the same treatment is, however, equally applied to flowering plants and grasses which would lend themselves far more easily to the European method of grouping. The reason for the peculiarity of treatment noticeable in these flower arrangements may rather be sought in the Japanese manner of observing and enjoying floral nature. Whereas the Western amateur devotes his attention mainly to the blossoms, the Japanese lover of flowers extends his admiration to every striking feature of the plant or tree producing them. The rugged nature of the plum trunk, with its straight, stiff shoots, or the graceful sweep of the branches of the weeping cherry, are to him inseparably associated with any beauty which the blossoms themselves possess. The lines of branch and stem, the form and different surfaces of leaves, and the distribution of buds and blossoms, all receive an equal share of attention and all play their allotted parts in designs. It may be said that the art under consideration is based upon a representation, more or less conventional, of *floral growth;* and, principally for this reason, the compositions are made to assume an open character in which the forms of branches, stems, leaves, and flowers are all clearly and individually expressed.

The vernacular term *hana*, translatable as *flower*, has, in the art of floral arrangement, a much wider signification than its nearest English equivalent. Among the so-called *flowers* of the seasons are included certain evergreens and other flowerless shrubs and trees, some of these holding very high floral rank. The pine and bamboo, for example, both occupy a very important place in what are called *flower arrangements*; also the maple with its reddening leaves is used as one of the principal *flowers* of Autumn.

In the choice of material, seasonableness is one of the principal points kept in view. The luxurious taste for choiceness, as implying rarity, is diametrically opposed to the rules of the art under consideration. Flowers blooming before or after their proper season are, with very few exceptions, rejected for Japanese floral compositions, such designs being in a manner intended to express the particular period of the year. April blossoms, used in any other month, would appear to the flower-artist as incongruous and out of place as winter clothing worn in summer-time. It therefore naturally follows that a proper cultivation of the floral art demands a thorough acquaintance with the nature and growth of all trees and plants employed: and, in the case of those common to several months, a close observation of the varying characteristics of the same plant during different seasons. The flag or iris, for example, which is common to different months of Spring, Summer, and Autumn, has a peculiar bend and vitality in its leaves, and a different length and vigour in its flower stems, during the various periods of its growth; and these distinctions are all kept in view when this flower is employed in compositions.

The natural locality of production of trees and plants; whether lake or river bank, mountain, or moor; greatly influences the character of the design employed. To arrange a water plant in the same manner and with the same surroundings as a land plant would be considered a great violation of the rules of appropriateness. Not only are blossom-bearing trees and flowering plants treated as perfectly distinct in character, but minor divisions as to locality of production are often observed in both. Among plants a distinction is made between ordinary land plants, forest plants, mountain plants, and water plants; and among trees, land trees, forest trees, and mountain trees are distinguished in certain cases.

The necessity for a proper familiarity with the nature of all flowers used in compositions is one reason strongly urged against the employment of rare or little known plants, however beautiful they may be. The use of wild flowers, only known to the botanist, as well as rare foreign flowers the names of which are not familiar to ordinary folk, is prohibited, unless the artist has previously made himself perfectly acquainted with all the natural characteristics of such flowers. As one exponent of the art has quaintly expressed it, the artist must be thoroughly imbued with a sympathetic feeling for the character, habits, virtues, and weaknesses of the members of the floral kingdom from which he seeks his material, till he possesses almost the same love and tenderness for their qualities as for those of living beings.

Preliminary to a study of Japanese Floral Art it is necessary to have some acquaintance with the principal flowers employed. These flowers are enumerated in the following pages under the heads of the different months to which they belong. In consequence of such a classification many flowers common to several months are repeated. It must be remembered that according to the old calendar the commencement of the first month,—which was at the same time the beginning of the Japanese Spring,—was about thirty days later than the first of January. The adoption in late years of the Gregorian calendar has therefore rendered it impossible to conform at the present day to all the rules laid down for the selection of flowers for special occasions. Such of the old fête-days as are now observed, are pushed back one month or more in time, and the flowers originally fixed as appropriate for their celebration are often unavailable, or recourse has to be made to premature or forced specimens. The following classification is according to the old calendar, existing when the whole theory of the art in question was established.

FLOWERS ACCORDING TO THEIR MONTHS

(OLD CALENDAR).

Against the Japanese names in the following list of flowers certain distinguishing signs are placed :—

* Stands for those trees or plants which are termed *Living Flowers*, being particularly characteristic of the month under which they are placed. Such flowers are much prized for felicitous occasions.

† Distinguishes the *Early Flowerings*,—flowers which are in advance of their proper season in the month under which they are placed. These have also their appropriate use in floral compositions.

‡ Indicates what are called *Passed Flowers*, and § stands for what are termed *Dead Flowers*. These names have reference to flowers which are late or passed in month or season, belonging properly to earlier months. The use of such flowers is forbidden for most ceremonial occasions. There exists also the term *Vulgar Flowers*, applied to wild plants, or to those of very common character which possess no fancy name; and the employment of flowers included under this head is not permitted except in the hands of the most experienced professors of the art. The use of cereals is also to be avoided.

FIRST MONTH (PRESENT FEBRUARY).

Japanese Name.	Botanical Name.	English popular Name.
*Fukujuso	Adonis amurensis	
§Suisen	Narcissus tazetta	Narcissus
*Uguisuso	Lithospermum zollingeri	
*Hakubai	Prunus mume	White plum
§Yanagi	Salix japonica	Willow
§Kan-giku	Pyrethrum sinense	Winter chrysanthemum.
*Yabukoji	Ardisia japonica	
*Rengio	Forsythia suspensa	
Tsubaki	Camellia japonica	Camellia
*Murozaki-Momo		Hot-house peach
*Obai	Jasminum sieboldianum	
*Kinsenkwa	Calendula officinalis	
*Choshun	Rosa indica	
*Mansaku	Hamamelis japonica	
§Robai	Chimonanthus fragrans	Chinese plum

SECOND MONTH (PRESENT MARCH).

Japanese Name.	Botanical Name.	English Popular Name.	Japanese Name.	Botanical Name.	English Popular Name.
‡Haku-bai	Prunus mume flore alba	White plum	*Shakunage	Rhododendron metternichii	Azalea
‡Hi-to		Red peach	†Tsutsuji	Rhododendron indicum	Azalea
‡Obai	Jasminum sieboldianum		†Kaido	Pyrus spectabilis	
*Ko-bai	Prunus mume flore rosa	Red plum	‡Obai	Jasminum sieboldianum	
*Higan-zakura	Prunus subhirtella		*Baran	Aspidistra lurida	
*Usu-to	Prunus persica	Pale peach	*Bijinso	Papaver rhoeas	
*Niwatoko	Sambucus racemosa		‡Uguisuso	Lithospermum zollingeri	
*Kemanso	Dicentra spectabilis		*Yukiwariso	Anemone hepatica	
†Azuma-giku	Erigeron thunbergii		*Kara-omodaka	Alisma plantago	
‡Kinsenkwa	Calendula officinalis		*Shunran	Cymbidium virens	
†Haru-giku	Chrysanthemum coronarium	Spring chrysanthemum	*Kobushi	Magnolia kobus	Magnolia
			§Tsubaki	Camellia japonica	Camellia
*Hotei-chiku	Bambusa sterilis	Bamboo	†Nashi	Pyrus ussuriensis	Pear
*Tennansho	Arisaema japonicum		*Sumomo	Pyrus triflora	
†Oka-kohone	Nuphar japonicum		*Ringo	Pyrus malus	Apple
*Rengio	Forsythia suspensa		*Asebo	Andromeda japonica	
*Anzu	Prunus armeniaca	Apricot	*Sendaihagi	Thermopsis fabacea	
†Hitoye-zakura	Prunus pseudo-cerasus	Single cherry	*Korai-giku		Corean chrysanthemum
†Yamabuki	Kerria japonica				
*Hitsujiso	Nymphaea tetragona		†Boke	Pyrus japonica	
*Niwa-ume	Prunus japonica	Garden plum	‡Mansaku	Hamamelis japonica	
*Wase-zakura	Prunus pseudo cerasus	Early cherry	*Choshun	Rosa indica	
*Niwa-zakura	Prunus pseudo-cerasus	Garden cherry	*Wasuregusa	Hemerocallis flava	
†Enishida	Cytisus scoparius		*Itadori	Polygonum cuspidatum	
†Mokurenge	Magnolia	Magnolia	*Sumire	Viola patrinii	Violet
†Suwo	Caesalpinia sappan		*Uikio	Foeniculum vulgare	
†Yobaikwa	Myrica rubra				

THIRD MONTH (PRESENT APRIL).

Japanese Name.	Botanical Name.	English Popular Name.	Japanese Name.	Botanical Name.	English Popular Name.
*Haku-to	Prunus persica flore alba	White peach	*Yamabuki	Kerria japonica	
*Usu-to		Pink peach	‡Rengio	Forsythia suspensa	
*Nippon-momo		Peach	Kobushi	Magnolia kobus	Magnolia
*Hi-to	Prunus persica	Red peach	Anzu	Prunus armeniaca	Apricot
*Noza-momo		Peach	Haru-giku	Chrysanthemum coronarium	Spring chrysanthemum.
*Hosumomo		Peach			
*Ko-to	Prunus persica flore rosa	Red peach	Jinchoge	Daphne odora	Daphne
*Gempei momo		Red and White peach	Ebine	Calanthe discolor	Kind of orchid
			Kome-zakura	Spiraea thunbergii	
‡Ri-to	Prunus triflora		Niwa-zakura	Prunus pseudo cerasus	Garden cherry
*Nashi	Pyrus ussuriensis	Pear	Suwo	Caesalpinia sappan	
*Ringo	Pyrus malus	Apple	Wase-zakura	Prunus pseudo-cerasus	Early cherry

[illegible] Name. [illegible]		English Popular Name	Japanese Name.	Botanical Name	English Popular Name
[illegible]		Korean chrysanthemum	†Kodemari	Spiraea cantoniensis	
[illegible]	[illegible]	Magnolia	*Ippatsu	Iris tectorum	Iris
[illegible]	[illegible]	Azalea	*Shaga	Iris japonica	Iris
	Myr[illegible]		†Botan	Paeonia moutan	Tree peony
[illegible]	[illegible]		*Awamoriso	Astilbe japonica	
[illegible]	[illegible]		*Kazaguruma	Clematis patens	Clematis
[illegible]			*Shiran	Bletia hyacinthina	
[illegible]	[illegible]	Peony	*Choshun	Rosa indica	
[illegible]	[illegible]	Iris	*Suzuran	Convallaria majalis	
*An[illegible]	Anemone [illegible]		*Tsurigane	Campanula punctata	Bluebell
*A[illegible]	[illegible]	[illegible]	Konnyaku	Conophallus konjak	
*Kaw[illegible]	[illegible]		Fuishi	Cytisus scoparius	
†T[illegible]	[illegible]	[illegible]	Kifuji	Wistaria chinensis	Yellow wistaria
‡W[illegible]	[illegible]		Fujimatsu	Larix leptolepis	
*[illegible]	[illegible]		Tampopo	Taraxacum officinale	Dandelion
[illegible]	Papaver [illegible]		Sumire	Viola patrinii	Violet

FOURTH MONTH (PRESENT MAY)

Japanese Name.	Botanical Name	English Popular Name	Japanese Name.	Botanical Name	English Popular Name
[illegible]	Chrys[illegible]	Chrysanthemum	*Natsuyuki	Deutzia sieboldiana	
Ku[illegible]	[illegible]		†Kirinso	Sedum kamtschaticum	
Sakaki	Cleyera japonica		*Hinageshi	Papaver rhoeas	
Azami	Cnicus japonicus		†Futoi	Scirpus lacustris	
[illegible]	Cont[illegible]		†Enishida	Cytisus scoparius	
[illegible]	Chenopodium album		†Hama-nadeshiko	Dianthus japonicus	
*Kakitsubata	Iris laevigata	Iris	‡Tsurigane	Campanula punctata	Bluebell
*Botan	Paeonia moutan	Tree peony	*Biyoyanagi	Hypericum chinense	
‡Ayame	Iris [illegible]	Iris	‡Kodemari	Spiraea cantoniensis	
§Ippatsu	Iris tectorum	I[illegible]	‡Suwo	Caesalpinia sappan	
*[illegible]	Rhodea japonica		§Kobushi	Magnolia kobus	Magnolia
*Kusa[illegible]	Althaea rosea		‡Shaga	Iris japonica	Iris
*Shakuyaku	Paeonia albiflora	Peony	‡Mokuren	Magnolia conspicua	Magnolia
[illegible]	Deutzia [illegible]		*Shimotsuke	Spiraea japonica	
*Yuri	Lilium	Lily	*Sendaihagi	Thermopsis fabacea	
†Zakuro	Punica granatum		‡Shiran	Bletia hyacinthina	
*Sendan	Melia azedarach		*Teppo-yuri	Lilium longiflorum	Lily
Tsutsuji	Rhododendron indicum	Azalea	*Sasa-yuri	Lilium japonicum	Lily
*Hanananten	Nandina domestica		*Hime-yuri	Lilium concolor	Lily
Kirishima	Rhododendron [illegible]		Hankwanso	Senecio japonicus	
‡Kinsenkwa	Calendula officinalis		*Giboshi	Funkia ovata	
†Kohone	Nuphar japonicum		*Kumagayeso	Cypripedium japonicum	
*Tessen	Clematis florida	Clematis	Atsumoriso	Cypripedium macranthum	

Japanese Name	Botanical Name	English Popular Name.	Japanese Name	Botanical Name	English Popular Name.
Benkeiso	Sedum erythrostictum		*Kobeku	Magnolia hypoleuca	Magnolia
Hanayu	Citrus aurantium		*Tampopo	Taraxacum officinale	Dandelion
Kikoku	Citrus fusca		*Fuji-kazura		
Shuro	Chamærops excelsa				

FIFTH MONTH (PRESENT JUNE).

Japanese Name	Botanical Name	English Popular Name.	Japanese Name	Botanical Name.	English Popular Name
*Kiku	Chrysanthemum coronarium	Chrysanthemum	‡Sasa-yuri	Lilium japonicum	Lily
*Giboshi	Funkia ovata		‡Hankwaiso	Senecio japonicus	
‡Shiran	Bletia hyacinthina		‡Benkeiso	Sedum erythrostictum	
*Kuchinashi	Gardenia florida		*Kuma-yanagi	Berchemia racemosa	
*Ukikusa	Lemna minor		*Kohone	Nuphar japonicum	
*Ajisai	Hydrangea hortensis	Hydrangea	*Sakaki	Cleyera japonica	
*Tochiso	Clintonia udensis		*Kodemari	Spiræa cantoniensis	
*Shimotsuke	Spiræa japonica		*Kwakuso	Phajus grandiflorus	
*Natsuyuki	Deutzia sieboldiana		*Futoi	Scirpus lacustris	
*Mokkokwa	Rosa banksiæ		*Hosoi	Juncus communis	
‡Hime-yuri	Lilium concolor	Lily	*Sankakui	Scirpus lacustris	
*Senno	Lychnis senno		*Kayatsurigusa	Cyperus iria	
*Matatabi	Actinidia polygama		*Sendan	Melia azedarach	
*Zakuro	Punica granatum		*Hanashobu	Iris lævigata	Iris
‡Biyoru	Hypericum chinense		*Kusa-ayame	Iris sibirica	Wild Iris
‡Hana-nanten	Nandina domestica		*Mankeishi	Vitex trifolia	
‡Tessen	Clematis florida	Clematis	*Nichinichiso	Vinca rosea	
‡Kirishima	Rhododendron obtusum	Azalea	*Kokwa	Carthamus tinctorius	
*Satsuki	Rhododendron macranthum	Azalea	‡Omoto	Rhodea japonica	
*Kirinso	Sedum kamtschaticum		‡Kakitsubata	Iris lævigata	Iris
*Natsuzukashi	Lilium thunbergianum	Lily	*Hakuchoke	Serissa fœtida	
*Kinginkwa	Goodyera parviflora		*Kwannonso	Reineckia carnea	
*Nadeshiko	Dianthus superbus		*Kurumi	Juglans regia	
*Kawara-nade-shiko	Dianthus superbus		*Ochi	Melia japonica	
‡Teppo-yuri	Lilium longiflorum	Lily	‡Kobeko	Magnolia hypoleuca	Magnolia

SIXTH MONTH (PRESENT JULY).

Japanese Name.	Botanical Name.	English Popular Name.	Japanese Name	Botanical Name	English Popular Name.
‡Kiku	Chrysanthemum coronarium	Chrysanthemum	‡Sekichiku	Dianthus chinensis	Kind of bamboo
‡Hana-nanten	Nandina domestica		*Hishi	Trapa bispinosa	
‡Omoto	Rhodea japonica		*Hakuchoke	Serissa fœtida	
*Ran		Orchid	‡Nadeshiko	Dianthus superbus	
‡Oshiroibana	Mirabilis jalapa				

Japanese Name.	Botanical Name.	English popular Name.
[illegible]	[illegible] speciosum	Lotus
[illegible]	Pardanthus chinensis	
[illegible]	Funkia ovata	
[illegible]	Pueraria thunbergiana	
[illegible]	Lysimachia clethroides	
*Kikio	Platycodon grandiflorum	
Tsuta	Vitis inconstans	Ivy
*Gampi	Lychnis grandiflora	
‡Senno	Lychnis senno	
*Kazaguruma	Clematis patens	Clematis
*Sakaki	Cleyera japonica	
*Mizu-aoi	Monochoria vaginalis	
Oguruma	Inula britannica	
*Mukuge	Hibiscus syriacus	
*Nozenka[illegible]	Tecoma grandiflora	
‡Kirinso	Sedum kamtschaticum	
‡Benkeiso	Sedum erythrostictum	
*Medo-hagi	Lespedeza sericea	
Asagao	Ipomaea hederacea	Morning glory
*Hirugao	Convolvulus japonicus	Convolvulus
*Yugao		Convolvulus
*Kohone	Nuphar japonicum	
†Kakitsubata	Iris laevigata	Iris

Japanese Name.	Botanical Name.	English popular Name.
*Kawara-nade-shiko	Dianthus superbus	
‡Futo-i	Scirpus lacustris	
‡Hoso-i	Juncus communis	
‡Sankaku-i	Scirpus lacustris	
*Ar[illegible]	Dianthus caryophyllus	
‡Shukaido	Begonia evansiana	
†Otogiriso	Hypericum erectum	
‡Omodaka	Alisma plantago	
‡Zakuro	Punica granatum	
‡Sanzashi	Crataegus cuneata	
*Manjushake	Nerine japonica	
*Sendan	Melia azedarach	
*Mankeishi	Vitex trifolia	
‡Kwannonso		
*Natsu-tsubaki	Stuartia pseudo-camellia	
*O-yuri	Lilium	Lily
‡Sasa-yuri	Lilium japonicum	Lily
‡Teppo-yuri	Lilium longiflorum	Lily
‡Hime-yuri	Lilium concolor	Lily
*Natsuzukashi	Lilium thunbergianum	Lily
*Itadori	Polygonum cuspidatum	

SEVENTH MONTH (PRESENT AUGUST).

Japanese Name	Botanical Name	English popular Name.
‡Kiku	Chrysanthemum coronarium	Chrysanthemum
*Kikio	Platycodon grandiflorum	
‡Ran		Orchid
*Gampi	Lychnis grandiflora	
‡Mukuge	Hibiscus syriacus	
‡Tsuta	Vitis inconstans	Ivy
*Sennichiko	Gomphrena globosa	
*Medo-hagi	Lespedeza sericea	
‡Hasu	Nelumbium speciosum	Lotus
‡Oguruma	Inula britannica	
*Senrio	Chloranthus brachystachys	
*Kuzu	Pueraria thunbergiana	
*Ominayeshi	Patrinia scabiosaefolia	
‡Asagao	Ipomaea hederacea	
*Hishi	Trapa bispinosa	
‡Yugao		Convolvulus

Japanese Name.	Botanical Name.	English popular Name.
‡Hirugao	Convolvulus japonicus	Convolvulus
*Hagi	Lespedeza bicolor	
‡Shukaido	Begonia evansiana	
‡Kohone	Nuphar japonicum	
‡Futo-i	Scirpus lacustris	
‡Hoso-i	Juncus communis	
‡Sankaku-i	Scirpus lacustris	
‡Mizu-aoi	Monochoria vaginalis	
‡Omodaka	Alisma plantago	
‡Otogiriso	Hypericum erectum	
*Shion	Aster tartaricus	Aster
*Keito	Celosia argentea	
‡Sawa-gikio	Lobelia sessilifolia	
*Hosenkwa	Impatiens balsamina	
*Fuyo	Hibiscus mutabilis	
*Ha-geito	Amaranthus melancholicus	

JAPANESE NAME.	BOTANICAL NAME.	ENGLISH POPULAR NAME.
‡Dandoku	Canna indica	
‡Hiogi	Pardanthus chinensis	
*Ukon	Curcuma longa	
*Kichijoso	Reineckia carnea	
*Kushide	Rhus semialata	
‡Kakitsubata	Iris laevigata	Iris
*Torikabuto	Aconitum fischeri	
‡Manjusake	Lycoris radiata	
‡Mankeishi	Vitex trifolia	
*Benkeiso	Sedum erythrostictum	

JAPANESE NAME.	BOTANICAL NAME.	ENGLISH POPULAR NAME.
‡Hakuchoke	Serissa foetida	
‡Kwannonso	Reineckia carnea	
*Rukiu-giku		Chrysanthemum
‡Aoi	Althaea rosea	
*Tsuru modoki	Celastrus articulatus	
‡Anji	Dianthus caryophyllus	
‡Nadeshiko	Dianthus superbus	Carnation
*Kawara nadeshiko	Dianthus superbus	

EIGHTH MONTH (PRESENT SEPTEMBER).

JAPANESE NAME.	BOTANICAL NAME.	ENGLISH POPULAR NAME.
‡Kiku	Chrysanthemum [illegible]	Chrysanthemum
‡Susuki	Eularia japonica	
‡Hasu	Nelumbium speciosum	
‡Tsuta	Vitis inconstans	Ivy
*Hagi	Lespedeza bicolor	
*Ogi		Kind of reed
‡Kakitsubata	Iris laevigata	Iris
*Shion	Aster tataricus	Aster
‡Yukinoshita	Saxifraga sarmentosa	
*Fujibakama	Eupatorium chinense	
*No-giku		Wild chrysanthemum.
‡Hosenkwa	Impatiens balsamina	
‡Fuyo	Hibiscus mutabilis	
*Himawari	Helianthus annuus	Sunflower
‡Keito	Celosia argentea	
‡Ominayeshi	Patrinia scabiosaefolia	
*Otokoyeshi	Patrinia scabiosaefolia alba	
*Torikabuto	Aconitum fischeri	
‡Tsuru-modoki	Celastrus articulatus	
*Ume modoki	Ilex sieboldi	
‡Mokuge	Hibiscus syriacus	
*Ganraiko	Amaranthus melancholicus	

JAPANESE NAME.	BOTANICAL NAME.	ENGLISH POPULAR NAME.
*Karukaya	Anthistiria arguens	
*Rindo	Gentiana scabra	
*Urazugusa		Kind of grass
*Hassaku-bai		Autumn plum
*Usumomiji	Acer palmatum	Kind of maple
‡Sanzashi	Crataegus cuneata	
*Hama-giku	Chrysanthemum nipponicum	
*Waremoko	Poterium officinale	
*Okina-gusa	Anemone cernua	
*Medo-hagi	Lespedeza sericea	
‡Sawa-gikyo	Lobelia sessilifolia	
*Sennichiso	Gomphrena globosa	
‡Mizu-aoi	Monochoria vaginalis	
‡Kohone	Nuphar japonicum	
*Nishikidama	Euonymus alatus	
‡Benkeiso	Sedum erythrostictum	
‡Kwannonso		
‡Ukon	Curcuma longa	
‡Kichijoso	Reineckia carnea	
*Kushide	Rhus semialata	
‡Riukiu giku		Chrysanthemum
‡Nishikigi	Euonymus alatus	

NINTH MONTH (PRESENT OCTOBER).

Japanese Name.	Botanical Name.	English Popular Name.
*Kiku	Chrysanthemum sinense	Chrysanthemum
*Nanten	Nandina domestica	
*Omoto	Rhodea japonica	
‡Hagi	Lespedeza bicolor	
‡Ume-modoki	Ilex sieboldi	
*Tsuru-modoki	Celastrus articulatus	
‡Ogi		Kind of reed
‡Rindo	Gentiana scabra	
†Suisen	Narcissus tazetta	Narcissus
*Susuki	Eularia japonica	
†Sawa-gikyo	Lobelia sessilifolia	
*Tsuwabuki	Senecio kaempferi	
*Cha-no-hana	Camellia theifera	Tea plant
*Yatsude	Fatsia japonica	
Sazanka	Camellia sasanqua	Camellia
*Tsuta	Vitis inconstans	Ivy
*Biwa	Photinia japonica	
*Shion	Aster tartaricus	Aster
*Kakitsubata	Iris laevigata	Iris
‡Karukaya	Anthistiria arguens	
‡Hama-giku	Chrysanthemum nipponicum	Chrysanthemum

Japanese Name.	Botanical Name.	English Popular Name.
*Sanzashi	Crataegus cuneata	
‡Ominayeshi	Patrinia scabiosaefolia	
‡Otokoyeshi	Patrinia scabiosaefolia [illegible]	
*Kochoka		
Shiki-zaki	General term for flowers blooming in all four seasons	
*Kabutogiku	Aconitum fischeri	
*Mizuhiki	Polygonum filiforme	
‡Fujibakama	Eupatorium chinense	
Yukinoshita	Saxifraga sarmentosa	
*Waremoko	Poterium officinale	
‡Moto-hagi	Lespedeza sericea	
*No-giku		Wild chrysanthemum
*Uzuragusa		Kind of grass
‡Riukiu-giku		Chrysanthemum
‡Kohone	Nuphar japonicum	
†*Koyo-moku*	General term for trees the leaves of which redden in the Autumn	
†*Yanagi-no-rui*	Trees of the willow kind	
*Nishikigi	Euonymus alatus	

TENTH MONTH (PRESENT NOVEMBER).

Japanese Name.	Botanical Name.	English Popular Name.
§Zan-giku		Late chrysanthemum
†Suisen	Narcissus tazetta	Narcissus
*Kan-giku	Pyrethrum sinense	Winter chrysanthemum
§Sanzashi	Crataegus cuneata	
*Cha-no-hana	Camellia theifera	Tea plant
*Biwa	Photinia japonica	
§Nanten	Nandina domestica	
*Omoto	Rhodea japonica	
*Neko-yanagi	Salix brachystachys	Kind of willow
§Tsuwabuki	Senecio kaempferi	
Kochoka		

Japanese Name.	Botanical Name.	English Popular Name.
**Shiki-zaki*	General term for flowers blooming in all four seasons	
**Koyo-moku*	General term for trees the leaves of which turn red in the Autumn	
*Nebukaso	Allium fistulosum	
*Yukinoshita	Saxifraga sarmentosa	
*Yatsude	Fatsia japonica	
*Karukaya	Anthistiria arguens	
‡Rindo	Gentiana scabra	
†*Hayazaki*-Tsubaki	Camellia japonica	Early camellia
†Toji-bai	Prunus mume	Early plum
*Jugwatsu-zakura	Prunus pseudo cerasus	Tenth-month cherry

ELEVENTH MONTH (PRESENT DECEMBER).

Japanese Name.	Botanical Name.	English Popular Name.	Japanese Name.	Botanical Name.	English Popular Name.
*Kan-giku		Winter chrysanthemum	**Ko[illegible]*	General term for trees the leaves of which redden in the Autumn	
*Suisen	Narcissus tazetta	Narcissus	‡Kachoke		
*Nanten	Nandina domestica		†Kan-botan	Paeonia moutan	Winter peony
§Omoto	Rhodea japonica		‡Sazankwa	Camellia sasanqua	Camellia
‡Neko-yanagi	Salix brachystachys		‡Yatsude	Fatsia japonica	
‡Toji-bai	Prunus mume	Early plum	‡Tsubaki	Camellia japonica	Camellia
‡Jugwatsu-zakura	Prunus pseudo cerasus	Tenth month cherry	**Shiki-zaki*	General term for flowers blooming in all four seasons	
‡Biwa	Photinia japonica				

TWELFTH MONTH (PRESENT JANUARY).

Japanese Name.	Botanical Name.	English Popular Name.	Japanese Name.	Botanical Name.	English Popular Name.
†Kangiku		Winter chrysanthemum	‡*Yanagi-no-eda*	Various kind of willow	
‡Suisen	Narcissus tazetta	Narcissus	‡Ro-bai	Chimonanthus fragrans	
*Kan-botan	Paeonia moutan	Winter peony	†Kinsenkwa	Calendula officinalis	
‡Nanten	Nandina domestica		§Kocho ke		
‡Omoto	Rhodea japonica		‡*Koyo-no-eda*	Various trees the leaves of which redden in the Autumn	
†Murozaki momo		Forced peach	‡*Shiki-zaki*	Flowers blooming in all four seasons	
†Haku-bai	Prunus mume flore albo		†Rengio	Forsythia suspensa	
*Tsubaki	Camellia japonica	Camellia			

Considerations of good or evil luck enter largely into the choice of flowers, especially when employed as decorations for occasions of rejoicing; and there are certain flowering plants and trees reputed to possess poisonous properties in their roots, stems, leaves, or blossoms, which are objected to at any time, their employment being considered unlucky and ominous. The following is a list of the principal of such poisonous flowers:—

OMINOUS FLOWERS.

Japanese Name.	Botanical Name.	Note	Japanese Name.	Botanical Name.	Note
Mochi-tsutsuji	Rhododendron ledifolium	The white flower species is not poisonous	Asebo	Andromeda japonica	Stem poisonous
Yaye-kwanzo	Hemerocallis fulva	Single flower species not poisonous	Yamagobo	Rhapanticum atriplicifolium	Red kind poisonous
Manjushage	Nerine japonica	Leaves are poisonous	Torikabuto	Aconitum fischeri	Root poisonous
Hana-suwo	Cercis chinensis	Flower poisonous	Karasungi	Rumex aquaticus	Stem said to be poisonous
Gojikwa	Pentapetes phoenicea	Highly poisonous	Inu-kusu	Machilus thunbergii	Root poisonous
Noren-kazura	Tecoma grandiflora	Tendrils poisonous	Tachimachi gusa	Aconitum lycoctonum	Very poisonous
Yama-ajisai	Hydrangea hirta	Root poisonous	Giboshi	Funkia ovata	Flowers poisonous
Hosenkwa	Impatiens balsamina	Leaves poisonous	Yama-nasubi	Datura alba	Very poisonous
Miyama-shikimi	Skimmia japonica	Leaves poisonous	Konnyaku	Arisaema japonica	Root poisonous
Yatsude	Fatsia japonica	Root poisonous	Kusagi	Clerodendron trichotomum	Leaves poisonous

In addition to the last named, all flowers having a powerful odour are considered unsuitable for placing before guests.

Among the flowers peculiar to the different months previously classified, some are considered specially appropriate for displaying upon fête days whilst others, though allowed at ordinary times, are interdicted for such important occasions.

As most of these flowers are to be found enumerated in the complete tables already given, the following classification is abbreviated, merely giving the Japanese names and the corresponding name in English, botanical names being printed only where no popular equivalent exists. It may be observed that this list includes in all twenty-four species of plants and trees, or, if different species of the same *genera* be classed together, the number of specially honoured flowers becomes reduced to fifteen:—

FLOWERS SUITABLE FOR FELICITOUS OCCASIONS.

FIRST MONTH (PRESENT FEBRUARY)

Fukujuso	*Adonis amurensis*
Yabukoji	*Ardisia japonica*
Haku-bai	White plum
Yanagi	Willow
Omoto	*Rhodea japonica*
Choshun	*Rosa indica*
Sho chiku bai	Combination of pine, bamboo, red plum

SECOND MONTH (PRESENT MARCH)

Momo	Peach
Yanagi	Willow
Kobai	Red plum
Omoto	*Rhodea japonica*
Haru-giku	Spring chrysanthemum
Choshun	*Rosa indica*

THIRD MONTH (PRESENT APRIL).

Sakura	Cherry
Momo	Peach
Haru-giku	Spring chrysanthemum
Omoto	*Rhodea japonica*
Choshun	*Rosa indica*

FOURTH MONTH (PRESENT MAY).

Botan	Tree peony
Shakuyaku	Peony
Moso chiku	Bamboo
Omoto	*Rhodea japonica*
Choshun	*Rosa indica*
Kiku	Chrysanthemum

FIFTH MONTH (PRESENT JUNE).

Kiku	Chrysanthemum
Omoto	*Rhodea japonica*
Moso chiku	Bamboo
Choshun	*Rosa indica*

SIXTH MONTH (PRESENT JULY).

Kiku	Chrysanthemum
Omoto	*Rhodea japonica*
Nanten	*Nandina domestica*
Choshun	*Rosa indica*

SEVENTH MONTH (PRESENT AUGUST).

Kiku	Chrysanthemum
Omoto	*Rhodea japonica*
Choshun	*Rosa indica*
Nanten	*Nandina domestica*

EIGHTH MONTH (PRESENT SEPTEMBER).

Kiku	Chrysanthemum
Omoto	*Rhodea japonica*
Hassaku-bai	Autumn plum
Nanten	*Nandina domestica*
Choshun	*Rosa indica*

NINTH MONTH (PRESENT OCTOBER)

Kiku	Chrysanthemum
Omoto	*Rhodea japonica*
Nanten	*Nandina domestica*
Choshun	*Rosa indica*
Suisen	Narcissus
Yanagi	Willow

TENTH MONTH (PRESENT NOVEMBER).

Zan-giku	Chrysanthemum
Suisen	Narcissus
Omoto	*Rhodea japonica*
Nanten	*Nandina domestica*
Choshun	*Rosa indica*
Yanagi	Willow

ELEVENTH MONTH (PRESENT DECEMBER)

Suisen	Narcissus
Kan-giku	Winter chrysanthemum
Omoto	*Rhodea japonica*
Yanagi	Willow
Nanten	*Nandina domestica*
Toji-bai	Early plum
Choshun	*Rosa indica*
Yaye-tsubaki	Double camellia

TWELFTH MONTH (PRESENT JANUARY)

Suisen	Narcissus
Kan-giku	Winter chrysanthemum
Yanagi	Willow
Omoto	*Rhodea japonica*
Haku-bai	White plum
Murozaki-momo	Forced peach
Choshun	*Rosa indica*
Tsubaki	Camellia

The following list of flowers, the use of which is prohibited for special occasions of ceremony or congratulation, is arranged without regard to the months to which they belong. The employment of such flowers is deprecated at any season, without reference to any particular month, unless no other flowers can possibly be obtained. The reasons for their rejection are not always very clearly defined; sometimes the objection is to the form, sometimes to the colour, occasionally to some supposed poisonous property, and often to what [illegible] little [illegible] than traditional superstition or caprice. It is not surprising, therefore, to find these objections disregarded at times by certain masters:—

FLOWERS PROHIBITED FOR FELICITOUS OCCASIONS.

[illegible]	*Eulalia japonica*	Nashi	*Pyrus [illegible]*
[illegible]	Aster	Kanzo	*Hemerocallis [illegible]*
[illegible]	*[illegible]*	Fuyo	*Hibiscus [illegible]*
[illegible]	*[illegible]*	Renge	Lotus
[illegible]	*Dianthus [illegible]*	Hototogisu	*Tricyrtis [illegible]*
[illegible]	Kind of reed	Mokuren	Magnolia
[illegible]	*[illegible]*	[illegible]	*Camellia [illegible]*
[illegible]	*[illegible]*	Ran	Orchid
[illegible]	Violet	[illegible]	*[illegible]*
[illegible]	*[illegible]*	Rengyo	*Forsythia [illegible]*
[illegible]	*[illegible] japonica*	Yoshi	*Phragmites communis*
[illegible]	*[illegible]*	Ashi	*Phragmites communis*
[illegible]	Chinese bamboo	Rindo	*Gentiana [illegible]*
[illegible]	[illegible]	[illegible]	*Gnaphalium [illegible]*
[illegible]	Large leaf [illegible]	Shakunage	Rhododendron
[illegible]	*[illegible]*	Kuchinashi	*Gardenia florida*
[illegible]	*Daphne [illegible]*	Asagao	*Ipomoea [illegible]*
[illegible]	*[illegible]*	[illegible]	*Funkia [illegible]*
[illegible]	*[illegible]*	[illegible]	*Pardanthus [illegible]*
[illegible]	*[illegible]*	[illegible]	*Ilex [illegible]*
Mokuge	*[illegible]*	[illegible]	*Smilax [illegible]*
[illegible]	*[illegible]*	Kohone	*Nuphar japonicum*
[illegible]	*[illegible]*	Hinoki	*Thuya obtusa*
[illegible]	*[illegible]*	Yatsude	*Fatsia japonica*
Keshi	*[illegible]*	Ajisai	*Hydrangea hortensis*

Ranking highest in the above list of felicitous flowers, the following seven are considered as *par excellence* those for ceremonies and congratulatory occasions:—

The *Kiku*, or Chrysanthemum, to which is given the fancy name *Choju-so*, meaning *Long-lasting Plant*, on account of its growing through all the four seasons.

The *Suisen*, or Narcissus, called by the fancy name of *Inyo-so*, or *Plant of the Two Sexes*, a name given to this flower because it comes in the Winter and lasts till the Spring of the following year.

The *Momiji*, or Maple, fancifully called *Dokuge-so* or *Poison-dispelling Plant*, because there is a popular superstition that it absorbs all poison and infection from the air.

The *Sakura*, or Cherry, regarded in Japan as the king of flowers.

The *Botan*, or Tree Peony, fancifully named *Fuki-gusa*, meaning *Plant of Wealth and High Rank*. The peony is said to be the queen of flowers in China.

The *Omoto*, or Rhodea japonica, much honoured because, unaffected by heat or cold, its leaf remains strong and green throughout the year.

The *Fuji*, or Wistaria, fancifully called *Niki-so*, meaning *Plant of the Two Seasons*, because, appearing between the third and fourth months, it belongs both to Spring and Summer. Though much honoured and used for felicitous occasions, the wistaria must not be employed at weddings on account of its purple hue, this being associated with mourning.

In addition to the above seven flowers, the *Kakitsubata* (Iris lævigata) also takes high rank, but on account of its purple colour, like the wistaria, it is prohibited for wedding ceremonies.

Hitherto attention has been directed to the principal flowering plants and trees of the country, and to the degree of esteem in which they are individually regarded, especially with reference to particular months. Many floral compositions consist of combinations of two or more different kinds of growth. The manner in which different species are combined is best explained when the whole theory of the Japanese methods of arrangement is discussed in a later chapter. But, apart from the manner of grouping them, there are certain prejudices in favour of and against different combinations of material which require mention in the present context. The following is a list of a few of the suitable and unsuitable combinations of flowers :—

APPROPRIATE COMBINATIONS.

Matsu (Pine)	with *Chōshun* (Rosa indica)	*Ume-modoki* (Ilex sieboldi)	with *Suisen* (Narcissus)
Matsu (Pine)	with *Kiku* (Chrysanthemum)	*Haran* (Orchid)	with *Nadeshiko* (Dianthus superbus)
Yanagi (Willow)	with *Suisen* (Narcissus)	*Tokusa* (Equisetum hyemale)	with *Sennō* (Lychnis senno)
Momiji (Maple)	with *Kiku* (Chrysanthemum, white or yellow)	*Hakubai* (White Plum)	with *Kinsenkwa* (Calendula officinalis)
Tsubaki (Camellia)	with *Suisen* (Narcissus)		

[illegible]	with [illegible] ([illegible] japonica)	*Take* (Bamboo)	with *Asa*[illegible] (Morning glory)
[illegible]	with [illegible]	*Nanten* (Nandina domestica)	with *Shi*[illegible] (White chrysanthemum) or *Suisen* (Narcissus)
[illegible]			

OBJECTIONABLE COMBINATIONS

[illegible]	[illegible]		*Maki* (Podocarpus macrophylla)
[illegible]	[illegible] ([illegible] japonica)	*Momo* (Peach)	with *Sakura* (Cherry)
[illegible]	[illegible] (Willow)	*Maki* (Podocarpus macrophylla)	with *Ran* (Orchid)
[illegible] (Nandina [illegible])	[illegible]	*Hinoki* (Thuya obtusa)	with *Ran* (Orchid)
[illegible]	[illegible]	*Shaga* (Iris japonica)	with *Ran* (Orchid)
[illegible]	[illegible]		*Omoto* (Rhodea japonica) or
[illegible]	[illegible] (Lotus [illegible])		*Kohone* (Nuphar japonicum)

All of the above combinations, both good and bad, are of trees or plants which are in bloom during the same month. They are, therefore, combinations which are practicable without violating the rules as to seasonableness. Those which are classified as objectionable are so considered, therefore, for reasons other than that of seasonableness. Sometimes the objection is based upon too close a resemblance in form or colour; in other cases, similarity of species, or of locality of production, leading to redundancy of expression in the composition, is the deterring cause. The peach and the cherry, for example, being both flowering trees and somewhat similar in character, are not considered suitable in combination.

HISTORY AND THEORY.

BEFORE proceeding further with an explanation of the Floral Art, it may be of some interest to enquire into the origin of a cult so curiously unlike, in its methods, any other hitherto followed in Europe. Japanese historians claim for it an Indian and religious origin. The doctrines of Buddha, deprecating as they did the wanton sacrifice of animal life, are said to have suggested the gathering of flowers liable to rapid destruction in a tropical climate, and prolonging their vitality by careful preservation. The survival of such a theory would seem to show that some form of the art was first introduced into this country with the adoption of the Buddhist faith, both as a part of its ritual—flowers being placed before the Buddhist spirits,—and also to provide a pious pastime for the priests. The religion of Sakya Muni, as is well known, reached Japan through China in the sixth century, and certain Chinese priests are mentioned as the first teachers of the art of arranging flowers in Japan. It also appears that the earliest native practitioners in this country were famous priests, amongst whom Shotoku Taishi, son of the Emperor Yomei, and Meikei Shonin, are particularly mentioned.

These primitive flower compositions were, however, of a more accidental, and, at the same time, of a more extravagant character than those of the art as it became afterwards modified and developed. They partook more of the nature of a Western bunch or nosegay, being crowded in arrangement and miscellaneous in substance; they also lacked the severe conventionality of later methods. The style of composition adopted still survives under the name of *Rikkwa*, meaning Erect Flower Arrangement, and is used for flowers placed as religious ornaments or offerings before shrines and tombs and as a votive decoration at marriage ceremonies. An approach to symmetry was a governing feature in the most elaborate of such compositions. Branches of blossoming trees or foliage were employed, to form a vertical central mass; and other flowers or bunches of foliage were disposed on either side in balancing groups. The idea of imparting graceful and harmonious curves to the different lines of the composition was as yet only partially developed. Unlike the later and more refined flower arrangements, this early style was remarkable for the mixture of a variety of different material, as many as twelve or more species of plants and trees being employed in one design. The chief lines of a composition, generally seven in number, were formed of branches of different growth, some of which were in full leaf or flower and others purposely light and sparse in character. Large leaves of other plants were used at the base or connection of these various branches to hide their bareness, and careful attention was given to the bends and curves of these leaves so as to reveal their front and back surfaces in a well balanced contrast. The shape and disposal of the hollows or openings in a floral design received as much attention as the principal lines. Terms such as,—"valley," "grotto," and "perspective distance," were applied to these openings, the fancy that a natural landscape was represented being always kept up. Even in this comparatively ancient development of the art, the proportion which the floral composition held to the vessel which contained it was fixed by rule, a practice which was followed in all later arrangements.

A special branch of the *Rikkwa* style, which approached to a kind of miniature gardening, was applied to the ornamental use of thick stubs and branches of trees and water plants arranged in broad shallow vessels with an admixture of small rocks and stones. The intention of pourtraying and suggesting landscape, which to the Japanese is present in every class of floral composition, is, in this particular style, patent to every observer. This kind of arrangement, in a somewhat degenerate form, may be frequently seen at the present day in floral decorations for large public rooms, old lichen-covered branches of pine, plum, or maple trees being the favourite material.

The *Rikkwa* style possessed numerous rules and an elaborate nomenclature applied to the different members of a composition. Plates XI. to XXI. inclusive illustrate the *Rikkwa* style of flower arrangement. The later and more popular styles, which it is the principal object of this work to expound, adopted, to some extent, similar terms and theories in a simplified form.

The more modern development of the Floral Art was simultaneous with a great fancy for ornamental vases of various kinds, which sprang up under the patronage of the famous Regent Yoshimasa, at Higashi-yama. The impetus given to the manufacture of choice vessels by this Prince, chiefly owing to his inauguration of the Tea Ceremonies and other polite accomplishments, was accompanied by an austere refinement in the methods of arranging flowers. It was mainly with the object of adaptation to the Tea Ceremonial that the first modifications in the Flower Art took place, and the chief reformers were the

PLATE XI.

DIAGRAM OF *RIKKWA* LINES

STRAIGHT *RIKWA* ARRANGEMENT. *SHIN* STYLE.

STRAIGHT *RIKKWA* ARRANGEMENT, *GIO* STYLE.

STRAIGHT *RIKKWA* ARRANGEMENT, *SO* STYLE.

BENT *RIKKWA* ARRANGEMENT, *SHIN* STYLE.

PLATE XVII.

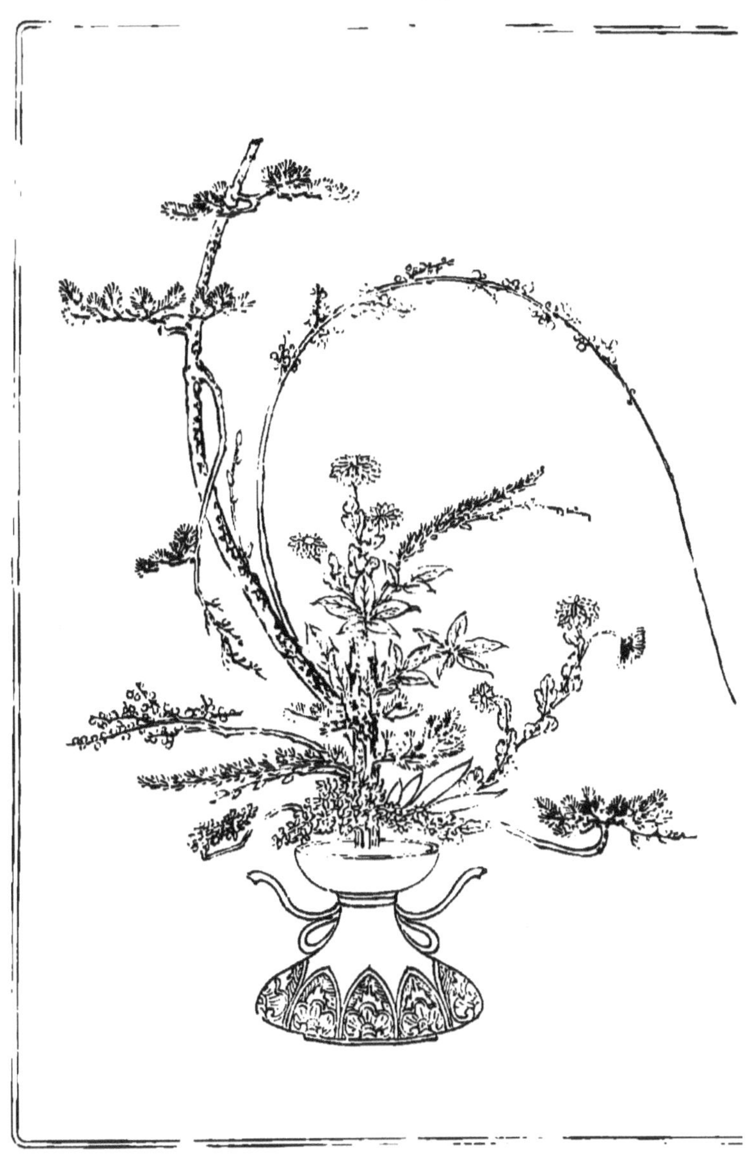

BENT *RIKKWA* ARRANGEMENT, *SO* STYLE.

RIKKWA DOUBLE STUMP ARRANGEMENT. (PINE, IRIS AND BAMBOO GRASS.)

RIKKWA COMBINED STUMP ARRANGEMENT FOR SHELVES.

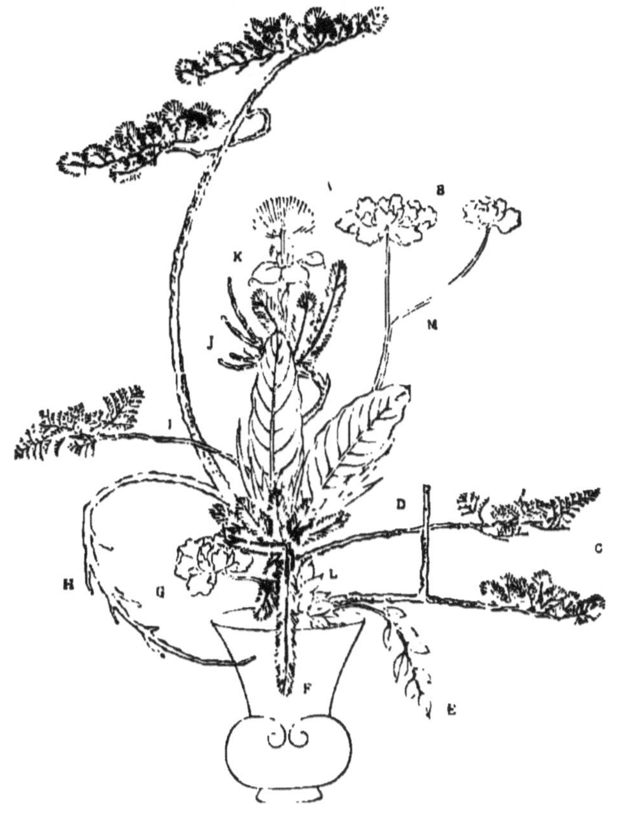
B
K
M
J
I
D
C
H
G
L
F
E

Chajin, or Professors of Tea. Sen no Rikiu, Senke, Sekishiu, and Enshiu, who were four of the principal flower designers of the Ashikaga and subsequent periods, were at the same time famous professors of the Tea Ceremonial. But the affected simplicity which ruled the Tea Room did not give full scope for the exercise of great elaboration in flower compositions, and other less austere forms of arrangement were developed, suited to the larger chambers of the nobility, but based upon the principles which had thus been inaugurated. A distinctive character and special proportion are given to flower compositions, as thus finally classified, according to the particular class of chamber which they adorn or the rank of the person in whose residence they are used.

The arranging of flowers has always been regarded in Japan as an occupation befitting learned men and *literati*, and though the ladies of the aristocracy have practised it, together with other polite arts, it is by no means considered as an effeminate accomplishment. Among its most enthusiastic followers appear the names of noted priests, philosophers, and even famous statesmen who have retired from public life.

Mixed up with the theory of the art, and imparting to it at first sight an air of quaintness and mystery, is a considerable amount of Chinese philosophy, together with many traditional superstitions. Ideas of good and evil luck control both the selection of material and the manner of arranging flowers for special occasions. Various virtues are attributed to the professors of the art, who are considered as belonging to a sort of aristocracy of talent, enjoying privileges of rank and precedence in society to which they are not by birth entitled. A religious spirit, self denial, gentleness, and forgetfulness of cares, are some of the excellencies said to follow the habitual practice of the art of arrangement of flowers. Philosophical classifications are resorted to for the purpose of distinguishing the different parts of floral designs. Thus Earth, Heaven, and Mankind are names employed in some styles to indicate the parts of a tri-lineal flower arrangement; Earth, Fire, Water, Metal, and Wood being used in the same way to designate the members of a five-lined design. Other Schools apply, in a similar manner, the names of abstract ideas, such as the five orders of Japanese versification, or the virtues of the human heart. The different methods of nomenclature are numerous, and would appear to have been adopted by the rival Schools principally with the object of imparting an appearance of originality and mystery to their own particular versions of what is practically one and the same art.

Again, the male and female principles so often referred to in Confucian philosophy are constantly applied to distinguish contrasting forms, surfaces, or colours in composition. It has ever been a favourite fancy of the Japanese to apply distinctions of sex to inanimate nature. In natural scenery, and landscape-gardening, it is customary to discriminate between *male* and *female* cascades, *male* and *female* plants and trees, and *male* and *female* rocks and stones. The distinction is not so much one of individual and separate quality as of forms placed in combination or contrast, and regarded as *male* or *female* in respect of one another. Thus the main torrent of a waterfall is considered *masculine*, and the lower fall in proximity *feminine*. In like manner, rocks used in gardening have no distinguishing sex, unless they are used in pairs or groups. In the case of two stones of different character placed side by side, the one of bolder and more vigorous shape will be called the *male*, and the other the *female* stone. Curious as such fancies may seem, they are of considerable value when applied in the arts of design, their observance helping to produce that harmony of well balanced contrasts which should pervade all artistic compositions. Nor are such ideas, indeed, quite foreign to certain branches of Western art. In architecture, for example, it is common to attribute *male* and *female* characteristics to the different orders of classic architecture.

In the Floral Art the idea of sex is applied to contrasting forms, long and short, large and small, angular and round, as well as to different kinds of growth, and to various colours in combinations. When a flowering tree is used in combination with a plant, the tree is considered as *male*, and the plant as *female*. With blossoms, buds are regarded as *female*, full flowers as *male*, and over-blown flowers again are classed as *female*; the time of full vigour receives the *male*, and the periods of weakness the *female* character. A similar fancy is applied to the different surfaces of leaves and to the different colours of flowers. Among colours, red, purple, pink, and variegated hues are *male*; and blue, yellow, and white are *female*. The front surface of a leaf is *male*, and its under surface is *female*. With flowers, the different forms of bud and full blossom, and in the case of leaves, the two surfaces, lend themselves easily to such distinctions; but berries seem to have defeated attempts at sexual classification, until Enshiu, one of the masters of the Flower Art, after observing the bulbul bird pecking at the fruit of a tree, devised the method of marking certain of the berries in a floral composition as if pecked by birds, and thus creating a dual character,—that of square and round. The *male* and *female* principles are also applied to the different directions of the stems of a composition, the right hand side and front of the design being regarded as *male*, and the left hand and further side as *female*. A stem on the left side of a composition turned to the front and back to the right is said to have *male* character, while to one on the right bent back to the left is attributed *female* character.

Beside sex, the different colours in flowers or leaves are said to have an order of *rank*. This idea of rank is applied principally to different colours of the same flowers. With most plants the white flower takes highest rank, but there are exceptions to this rule. The yellow chrysanthemum takes precedence of those of any other colour. With peach blossoms, the pale pink colour ranks first, though there exists a rich red specimen of great beauty and rarity. In the case of other flowers the colours taking the highest rank are with irises, purple, with camellias, red; with wistarias, lavender; with peonies, red, with valerians, yellow; with convolvuli, dark blue, with kerrias, yellow; with kikios, light purple; and with lespedezas, pink.

As before mentioned, a number of different Schools of Flower Arrangement sprang up after the impetus given to culture by the great art-patron, Yoshimasa. The principal styles of composition practised by these different Schools are known as the *Ko-Riu, Ikenobo-Riu Enshiu-Riu, Shinsho-Riu, Sekishiu-Riu, Misho-Riu, Senke-Riu, Yabunouchi-Riu, Kodo-Riu*, and *Seizan-Riu*. The masters of any of these Flower Schools would resent as ignorant the idea of confounding their theories with those of a rival style. Each School has its own special terms, philosophical mysteries, and secret degrees, but to enter closely into these peculiarities would be almost like discussing some of the more trivial differences between sects in the same religion. The main principles of the art are the same throughout, and the floral designs produced, regarded as works of art, are as similar as any desings can possibly be. The compositions of the *Rikkwa* and *Ikenobo* styles are the only ones which present a distinct character easily recognisable from those of other styles. In the present work the *Enshiu* style has been chiefly followed, this being the most elaborate and one of the most popular of the more modern styles, but numerous illustrations have been taken from compositions of other Schools.

The *Enshiu* style was originated by a retainer of the Shogun Iyeyasu, called Kobori Totomi no Kami. He was a distinguished professor of the Tea Ceremonial, and became teacher of this accomplishment to the Shogun's heir, Iyemitsu. Compared with some of the other styles, that of Enshiu is characterized by a greater degree of artificiality or artistic affectation, and this makes it specially adapted for the purposes of analysis for and a thorough explanation of the principles of the art.

LINEAL DISTRIBUTION.

AN analysis of Japanese flower arrangements shows that the lines or directions taken by the different stems or branches form the basis of all compositions. While European floral decorations' are merely combinations of masses of colour, in which blossoms and leaves alone play a part, those of Japan are synthetic designs in line, in which every individual stem, flower, and leaf stands out distinctly silhouetted. The treatment employed may be likened somewhat to the methods of distributing carved foliage in architectural panels as followed during the middle ages in Europe.

The surface of the water in which the flowers are placed is technically considered to be the soil from which the floral growth springs; and the designer must here convey the impression of stability and strength. However good the upper lines of the composition may be, a weak *springing* at the base deprives them of life and vigour: for it must be remembered that not flowers alone are displayed, but floral growth and vitality are to be expressed in the designs. The direction of the stems at starting need not be strictly vertical, but, if curved, the curves employed should be strong ones, and all weak bends and angles should be avoided. As a composition consists of several main lines, it follows that there will be several lines of *springing*, or origin. In some cases, the *springing* lines are all united from below the surface of the water to a point some distance above, after which they separate in tangental curves in different directions; in other cases, each stem-line is kept distinct throughout, being separated from the others immediately from the point of origin.

In the distribution of the principal lines of the composition from the point of their separation, the artist studiously avoids an equal-sided or symmetrical arrangement, but he obtains a balance of a more subtle kind, which is at the same time productive of a pleasing variety of form. Balance and harmony without repetition is a governing principle in this as in other Japanese arts. The lines of each stem, or, in cases where numerous slender stems are combined, the central lines of each group of stems, receive first attention.

The triple arrangement—by which is meant that governed by three prevailing lines,—may be taken as the original model for all arrangements (see Plate XXII A.).

The three lines of such compositions may with sufficient fidelity to the more quaint native nomenclature, be called, *Principal*, *Secondary*, and *Tertiary*. The *Principal*, as its name implies, is the central and longest line of the design, and this is made to form a double curve with the upper and lower extremities nearly vertical and in a continuous line, the general shape thus assumed being that of an archer's bow. The *Secondary* should be about half, and the *Tertiary* about one quarter, of the length of the *Principal*, supposing all to be straightened out; and these two lines are arranged on different sides of the *Principal* in graceful double curves of varied character. As a general rule, the *Secondary* has a more vertical, and the *Tertiary* a more lateral tendency; the former being on the outside of the arched bow formed by the *Principal*, and the latter making a counterpoise on its hollow side. According as the hollow of the *Principal* faces right or left, the arrangement is called a right or left composition. By changing the direction and giving a different character to the curves of these three lines a great number of designs are produced. Some of these are shown in Plate XXII B., the more violent curves being favoured by the *Enshiu* School.

To produce a five-lined arrangement, two additional lines are introduced between the three previously named. The one placed between the *Principal* and the *Secondary* is called the *Support*, and that between the *Principal* and *Tertiary* is called the *Sub-principal*. The *Support*, in its length and value, approaches more to the *Secondary* than to the *Principal*: while the *Sub-principal*, as its name would imply, in size and importance approaches more the *Principal* than the *Tertiary*. In this way, it may be observed, a proper lineal balance and harmony is obtained. For supposing figures are used to indicate relative size and importance, and supposing the three elementary lines to be valued 4, 2, and 1, according to their respective lengths and degrees of importance, then, to preserve a proper balance with the addition of two extra lines, that introduced between 4 and 1 should be longer and nearer in value to 4 than that between 4 and 2, which should approximate more the dimensions and character of 2. These additional lines, besides having different intermediary lengths, have special curves given to them, and are arranged so as best to fill the intervals, with due regard to variety and balance. In the seven-lined arrangements, two more extra members are added, one called the *Side-line* and the other the *Trunk-line*. Their lengths are about intermediary between those on either side of them, the *Side-line* being placed between the *Support* and the *Tertiary*, and the *Trunk-line* between the *Sub-principal* and the *Secondary*.

The different members of the above lineal arrangements have curious fancy names

A

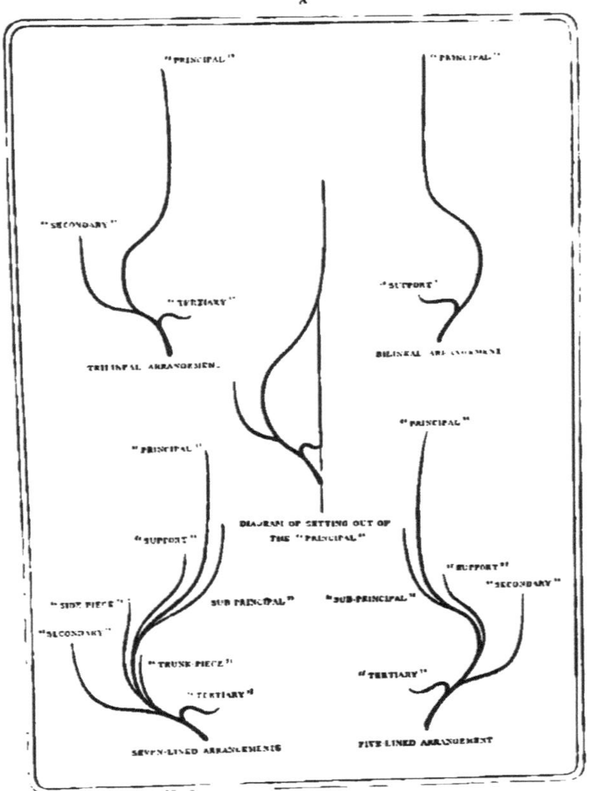
"PRINCIPAL"
"SECONDARY"
"TERTIARY"
"PRINCIPAL"
"SUPPORT"
DIAGRAM OF SETTING OUT OF THE "PRINCIPAL"
"PRINCIPAL"
"SUPPORT"
"SIDE PIECE"
"SECONDARY"
SUB PRINCIPAL"
"TRUNK PIECE"
"TERTIARY"
SEVEN-LINED ARRANGEMENTS
"PRINCIPAL"
"SUPPORT"
"SECONDARY"
"SUB-PRINCIPAL"
"TERTIARY"
FIVE LINED ARRANGEMENT

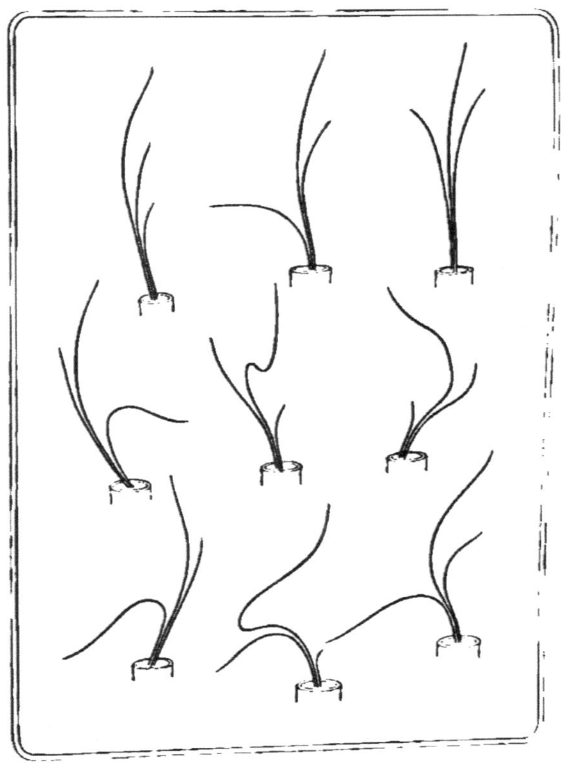

bestowed on them by the different Schools. For the triple style such classifications as that of *Father, Mother,* and *Self,* or that of *Heaven, Earth,* and *Mankind,* are used; and for the five-lined style the terms *Centre, North, South, East, West;* or *Earth, Fire, Water, Metal, Wood;* and sometimes *Yellow, Red, Black, White,* and *Blue* are employed. The *Enshin* School apply the following curious ideas to the different lines in an arrangement, namely:—*The Heart, Help, The Guest, Skill, The Finishing touch;* referring to the different sentiments which should inspire the designer in his treatment of the different lines. This School maintains that there are three secrets of arranging flowers. First, the art of giving expression and feeling to compositions; Second, the art of showing the character of the materials, whether plants or trees; Third, the art of keeping in mind the season and occasion of the arrangement by avoiding incongruous combinations.

The general form of the above groups of three, five, and seven lines depends mainly upon the amount of curvature given to the *Principal* or centre-most line. In the simpler and less affected styles, the bow-like curvature of the *Principal* is slight and strong, but in the more exaggerated compositions this governing line is boldly bent sideways from a point a few inches above the *springing,* and turned in a quick curve back again so as to bring the upper extremity vertically above the base, and to preserve the centre of gravity of the whole. The general form thus imparted is, as mentioned above, somewhat like that of a strung bow. From such a shape the transition is not great to that produced in the arc of a bow by drawing its string, a more violent lateral curvature with less vertical height being produced. This latter character is given to the *Principal* chiefly when used in positions where too much height would meet with actual or optical obstructions and be detrimental to a good effect.

Fig. 1.

Such treatment is often followed in floral arrangements placed below a shelf or in front of a hanging picture which must not be hidden. This modification in the character of the *Principal* necessitates corresponding changes in the direction and curvature of the other lines of the composition.

Up to the present, slight or violent as may

be the bow-like curvature imparted to the *Principal*, it has been always presumed that its general direction,—that is, the direction of a line joining its base and top,—is vertical. Thus, supposing a tri-lineal composition enclosed in a right-angled triangle, such triangle would be standing on its point, with the hypotenuse, corresponding with the *Principal*, placed vertically (see Fig. 1). There is, however, another style of design applied to a large class of flower arrangements in which the *Principal* line of the composition has a horizontal or almost horizontal direction; and, supposing a tri-lineal group of this style enclosed in a right-angled triangle, the hypotenuse of the triangle would lie horizontally (see Fig. 2). This latter style is chiefly applied to flowers arranged in hanging vessels, or in standing vases when placed on raised shelves, the intention of such compositions being to suggest floral growths on the edges of cliffs or banks which lean over laterally. In such elevated arrangements it often happens that one of the auxiliary lines on the side towards which the *Principal* bends over receives a decided droop and proportionate lengthening in order to emphasize the droop. Such drooping lines are technically called *Streamers*. The *Streamer* character may be imparted to either of the auxiliary lines, but redundancy must be avoided, and it is regarded as a fatal error to introduce *Streamers* on both sides of the *Principal*. The *Streamer* is very much used for floral arrangements in suspended vessels, in tall bamboo vases having mouths in their sides or in receptacles placed upon the edges of raised shelves. The prevailing idea in each of these cases is that the composition must suggest the wild growth on the edge of a bank or precipice over which the *Streamer* droops (see Plate XXIII. A).

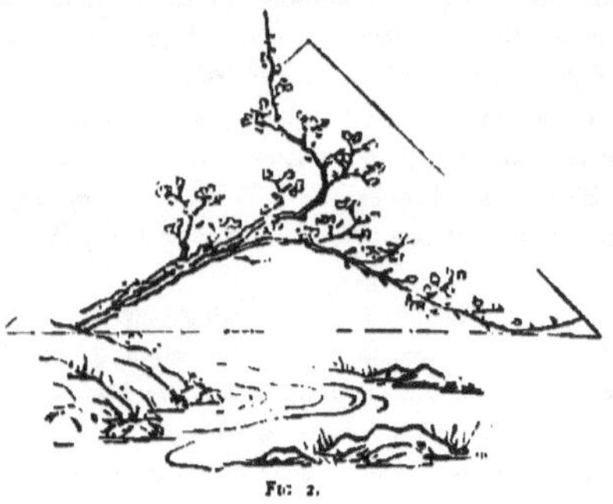

Fig. 2.

The above description has been confined to three, five, and seven-lined designs. Single line and double line compositions, as well as those of nine, eleven, or more lines, are sometimes made, but their use is very rare. The double-line form is employed occasionally for the simplest of all flower arrangements, namely, that of one flower and one leaf.

The different lines of a composition have hitherto been spoken of as if existing in one vertical plane parallel to the spectator, but actually, in addition to the directions

PLATE XXIII

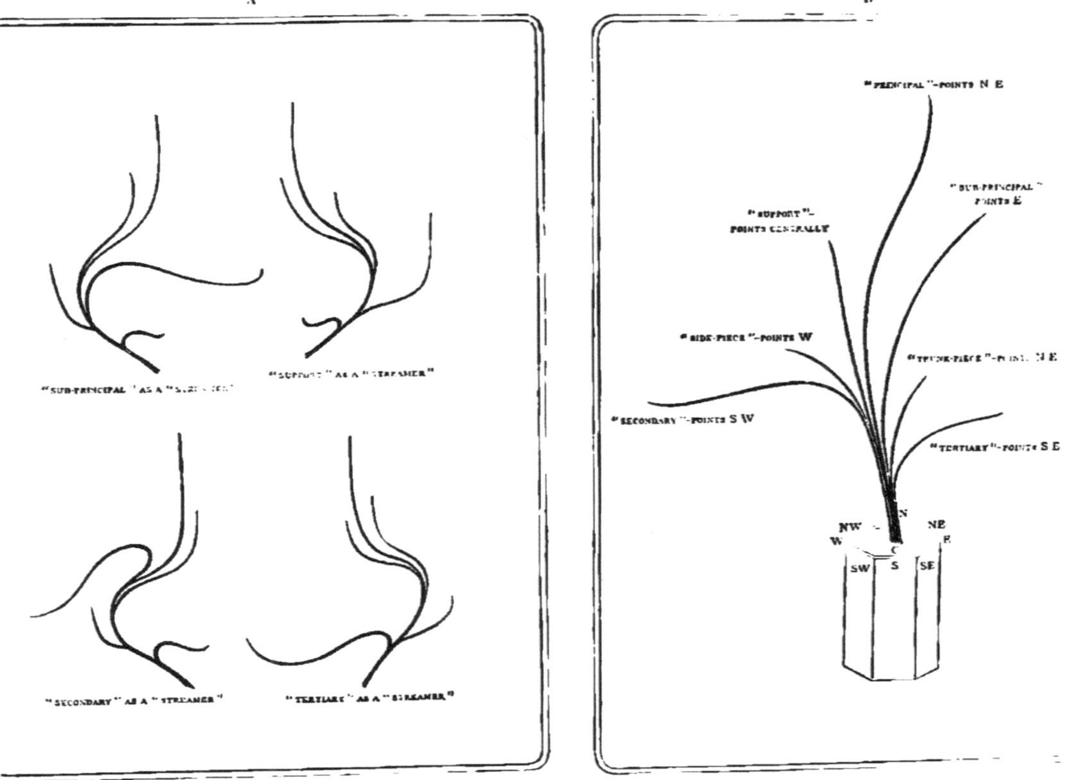
A
"SUB-PRINCIPAL" AS A "STREAMER"
"SUPPORT" AS A "STREAMER"
"SECONDARY" AS A "STREAMER"
"TERTIARY" AS A "STREAMER"
B
"PRINCIPAL"-POINTS N E
"SUB-PRINCIPAL" POINTS E
"SUPPORT"- POINTS CENTRALLY
"SIDE-PIECE"-POINTS W
"TRUNK-PIECE"-POINTS N E
"SECONDARY"-POINTS S W
"TERTIARY"-POINTS S E
N
NW
NE
W
E
C
SW
S
SE

mentioned to the right, left, upwards, and downwards, these lines have also directions of varying degrees forwards or backwards. In other words, their extreme points would require a solid and not a plane figure for their enclosure. These directions are best explained by supposing a bundle of stems placed in a vase of octagonal plan, and designating the eight different facets of the vessel respectively as north, north-east, east, south-east south, south-west, west, north-west: then, imagining the south face of the vessel immediately in front of the spectator, and bearing in mind that all the stems coincide for some little distance from their origin, the cardinal directions which they take after separation would be as follows (see Plate XXIII B):—The *Principal* bends north-east, the *Secondary* south-east, the *Tertiary* south-west, the *Sub-principal*, which is between the *Principal* and the *Tertiary*, turns east, and the *Support*, which lies between the *Principal* and *Secondary*, terminates centrally over the vase. Of the two extra lines used in seven-lined arrangements, the *Trunk-line* bends north-east, and the *Side-line* bends west. In this manner a pleasing variety back and front, as well as laterally, is given to the compositions, and they are not the flat arrangements which might at first sight be supposed from explanatory drawings. Though designed principally with the object of being seen from a point of view immediately in front, they obtain by the above treatment solidity and variety, and their effect when regarded from points to the right or the left are also taken into consideration.

Certain errors in arranging the lines of floral designs are pointed out to be strictly avoided. The first is that of *Cross-cutting*, which is produced by allowing two or more lines visually to intersect one another, so as to form angles.

A similar fault, called *View-cutting*, exists when a stem or twig cuts across its parent stem or branch; but this is allowed for certain trees, as for example the Plum, with which such inter-crossing is a characteristic feature of the growth.

Another error, called *Parallelism*, occurs when two or more adjacent stems or branches are exactly parallel to one another, a fault much aggravated when the lines are, in addition, of equal or nearly equal length.

Window-cutting, so named after the curved or cusped windows of the temples of the country, is the name given to an error occasioned by allowing stems to cross and re-cross in curves so as to form loop-like openings.

Lattice-cutting is an exaggerated form of *Cross-cutting*, and this defect occurs when several stems cross in parallel directions so as to suggest the appearance of lattice work.

Another fatal error is that of employing the *Double streamer*, or drooped branch, on both sides of the same composition. The above rules of prohibition, it will be observ-ed, are for the most part similar to those followed in floral and other ornamental designs in the best styles of the architectural arts, being intended to prevent harsh, conflicting, or redundant effects and to be productive of general harmony of line. They are illustrated in Plate XXIV.

The various directions of line imparted to the stems of plants and branches of trees on the above principles are obtained, first, by a careful selection of suitable material: then, by twisting, bending, building together, and fixing at the base: and, lastly, by means of cutting and clipping away defective and superfluous parts. Some special methods of manipulation will be afterwards considered.

PLATE XXIV

A

B

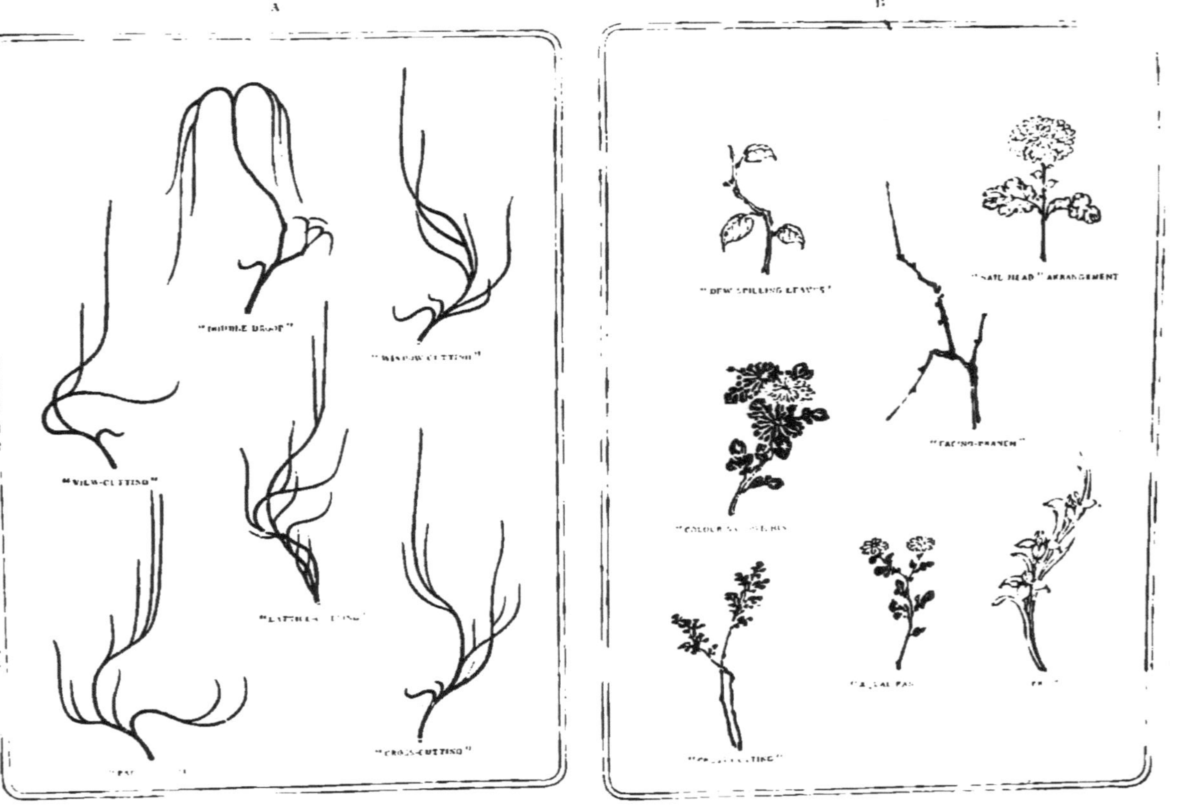

SELECTION OF MATERIAL.

UPON the general lines of composition already indicated, flower arrangements are made, sometimes with one species of tree or plant alone, and sometimes with a combination of two or more species. The use of many different kinds of flowers in one composition, though followed in the earlier styles, is opposed to the principles of the purer styles afterwards developed.

Combinations of two or three different species are, however, very common, and especially applied to arrangements in vessels having two or three openings. In all compositions, single or combined, the special nature of the different materials employed is carefully kept in mind, anything at all suggestive of the inappropriate being most scrupulously avoided. Important distinctions are made between trees and plants, and between land and water plants. The locality of production, whether mountain, moor, or river, considerably influences the arrangement adopted. Each flower has its proper month or season, and many flowers which are common to more than one season have special characteristics of growth, or of form, during the different seasons. These points of distinction are carefully studied, and are expressed in the artificial arrangements, within the limits of the art. Second flowerings, or flowers blooming out of their proper season, are discarded in flower compositions. As an example of this may be mentioned a late kind of peach blossom, appearing in the Summer, which, beautiful as it is, may not be employed, because the peach blossom is specially a flower of the Spring time.

In arranging two or more species in one composition, variety must be sought by combining branches of trees with plants. In the case of three lines being used, the branches of a tree should never be "supported" on both sides by a plant, nor should a plant be placed in the centre with a tree arrangement on either side. This fault is called by a term which will be better understood if freely translated as *sandwiching*. In a triple arrangement it is plain that two branches of the same kind of growth must be used, but the [illegible] adjoin, and not *sandwich* the remaning one. As an example of a defective arrangement may be taken a composition with irises (*plants*) in the centre, and branches of azalea and camellia (*trees*) on either side. A correct composition would be one with a plum branch (*tree*) in the centre, a pine branch (*tree*) on one side, and a bamboo stem (*plant*) on the other. For examples of the above correct and incorrect arrangements see Plate XXV. Some Schools allow occasionally the violation of this rule, and permit the *supporting* of a tree on either side by a plant, provided that the tree is a mountain tree and that one of the plants is a land and the other a valley or water plant.

The above rules were no doubt made principally to prevent the weak and insipid arrangements likely to be produced by the careless violation of the principles which they express, especially in the hands of the inexperienced. Plants, as before mentioned, are regarded as *female* with respect to trees, which are considered *male*, because the former are weaker and more graceful in character than the latter. A slender plant flanked on either side by tree branches would give a weakness of effect to the centre of the composition, and the reverse arrangement would give too much strength to the centre and weakness to the sides. In addition to this, such arrangements would have a more or less symmetrical character, and symmetry is disliked throughout the whole of the art under consideration. Like most arbitrary rules, such directions were often departed from by the more advanced professors, and there are even recognized exceptions which are universally admitted as correct. For example, two kinds of pine branches may be used together with a plum branch in a triple arrangement. Also in double compositions the willow and camellia are used together, and the *Celosia argentea* is often employed with the *Serissa fœtida*.

As previously stated, the branches and foliage of evergreens, and even of deciduous trees, are much used in floral compositions, the arrangement often being without a single blossom. It is, however, laid down as a general rule, that no flower-bearing plant is to be employed with leaves only, nor must plants or trees which bear leaves at blossom time be used with flowers alone. The following are exceptions to this rule:—The large leaved Chinese orchid has a flower, but it is very insignificant and grows below, and this plant is therefore technically treated as a flowerless one. The *Iris japonica* is sometimes arranged for its leaves only, before the flowers appear, and it then receives a special name. The leaves of the Summer narcissus, sometimes called by the fancy name of *Golden pillar plant*, are faded and withered before the flowers appear, and they may therefore be removed and discarded.

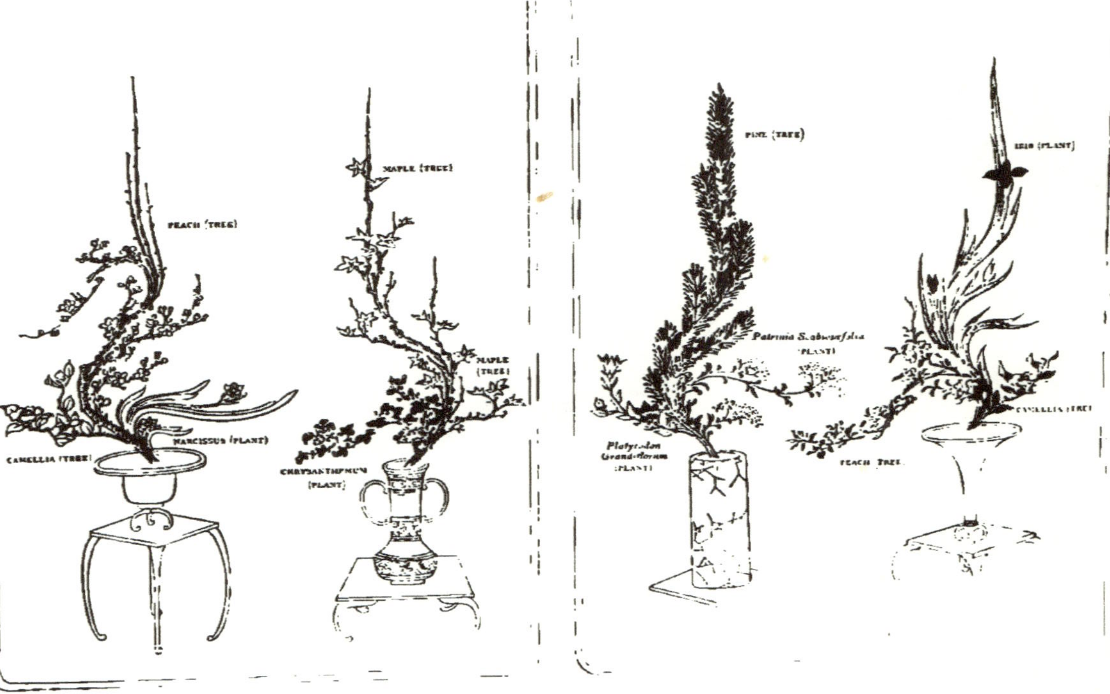
PEACH (TREE)
NARCISSUS (PLANT)
CAMELLIA (TREE)
MAPLE (TREE)
MAPLE (TREE)
CHRYSANTHEMUM (PLANT)
PINE (TREE)
Patrinia
(PLANT)
Platycodon Grandiflorum (PLANT)
IRIS (PLANT)
CAMELLIA (TREE)
PEACH TREE

All flower compositions must partake as much as possible of the spirit of the season in which they are used. Spring arrangements should be strong and powerful in line, like the growth of early vegetation: Summer arrangements must be full and spreading; whilst those of Autumn should be spare and lean; and those of Winter, withered and dreary.

Mention has been made above of certain fatal errors in combining the stem-lines of a flower composition, which are to be carefully avoided. There are other faults in arrangements which concern the distribution of leaves and flowers. Large blossoms should not be allowed to occur in regular steps or rows in a composition, and this error is called the *Stepped arrangement.* Another fault called the *Nail-head style*, after the stiff metal flower ornaments used to cover nails in Japanese buildings, consists in a flower turned exactly full view to the spectator with leaves on either side. The term *Colour-cutting* is given to the error of placing blossoms of the same colour on either side of a single blossom of another colour; as, for example, a white chrysanthemum between two red ones. This is only another example of the prejudice prevailing against *sandwiching* of any kind. The term *Dew-dropping leaf* is given in a composition to a leaf which droops down in a feeble manner so that it could not support a drop of water: this is greatly objected to. The above errors are illustrated in Plate XXIV.

Three distinct stages of vitality are observed both in flowers and leaves. In flowers, there are the full blossoms, the half open blossoms, and the buds: and in leaves, the young green leaf, the full leaf, and the reddening or falling leaf. In flower arrangements with one material, as for example with cherry or peach blossoms alone, a different character of blossom is selected for the chief lines of the composition. For the *Principal*, full blown flowers will be used; for the *Secondary*, half open flowers; and for the *Tertiary*, buds are employed. Some designers, on the principle that the half open flower is more powerful than the full blown blossom, use the half open flowers for the *Principal*, and the full blown blossoms for the *Secondary*. Straight leaves are considered strong, and curled or bent leaves weak: the strong flowers should be near the weak leaves, and the strong leaves should adjoin the buds or over-blown flowers. A flower below a leaf is weaker than one above. In thinning out leaves in a composition, two strong leaves must remain for every weak one.

A

B

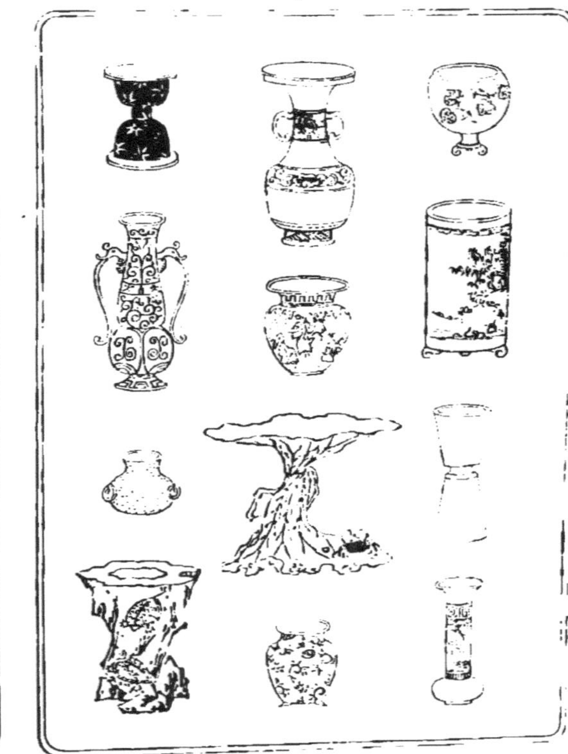

VARIOUS STANDING VASES OF BRONZE AND PORCELAIN

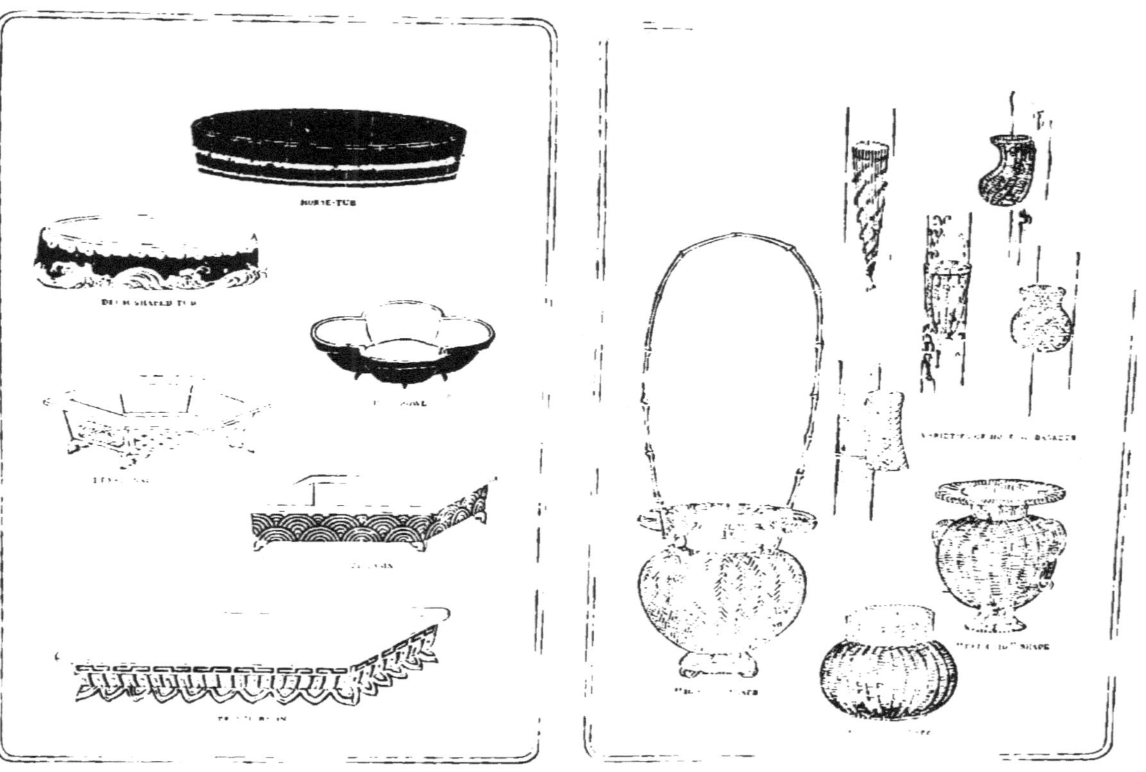
HORSE-TUB

FLOWER VESSELS.

STANDING VASES

HAVING classified the different plants and trees which supply material for the Japanese floral designer, it is necessary to describe the various sorts of flower vessels which are employed for receiving the compositions. The form and character of such vessels considerably influence the nature of the floral arrangements placed in them. These receptacles are of a diversity of shapes and of different materials, such as wood, porcelain, pottery, bronze, brass, iron, and basket-work. Without attempting to trace the development of the different art-industries as applied to vessels for receiving flowers, it will be necessary to refer to the various shapes, proportions, material, and decoration of such vessels, so far as they control or are governed by the floral arrangement adopted. The most ancient receptacle used was a long-necked earthenware or bronze vase of considerable height, and the flower composition placed therein was made proportionately tall. The difficulty of balancing such high arrangements led to the use of shorter and broader vessels and to a corresponding lowering of the compositions. This fact shows that from the earliest times a recognized connection of proportion existed between the floral group and the vessel in which it was placed.

Fig. 1

It is important to note that the surface of the water in which the flowers are placed is technically regarded as the soil from which the growth springs, and the designer must here convey the impression of a stable origin. With this principle in view, and guided also by the necessity previously mentioned of avoiding too great a height in flower composition a sort of broad-mouthed vase became the favourite form used. Of this kind there are many shapes, generally in bronze, and varying from a low saucer-like vessel to a trumpet-shaped form, supported sometimes on short legs, and sometimes upon ornamental castings representing such subjects as rocks, water, spray, or an animal group.

Other kinds of vessels, corresponding more to the ordinary vases employed in the West, having necks and oval or cylindrical bodies, are also much used. Among these, the vases with tall narrow necks are employed for the simplest arrangements, such as are displayed in the tiny tea rooms where rules of austerity necessitate very light and quiet floral compositions. Vases intended for elaborate flower arrangements are always necessarily of the wide-mouthed kind, to ensure an appearance of stability at the base of the flower stems. A few of these different vases are shown in Plate XXVI., but there exist a number of other shapes in common use, besides several fancy forms which are occasionally employed. With the ordinary tall vase, whether of wide or narrow mouth, the height of the flowers is generally fixed as approximately one and a half times that of the vase. Some Schools increase this proportion, giving double the height of the vase for the floral composition.

VESSELS FOR WATER PLANTS.

For the purpose of displaying, in a suitable and suggestive manner, water plants and grasses, low tub-like vessels came into use from quite early times. There are two principal kinds employed, called respectively the *Sand-bowl* and the *Horse-tub*. The *Sand-bowl* is a broad shallow vessel, oblong, polygonal, or sometimes oval in plan, made of bronze or porcelain, and containing a layer of pebbles or sand covered with water. The *Horse-tub* is, as its name implies, a shallow tub of wood, circular in plan, and generally lacquered black. Its use as a flower vessel is said to have originated during the exigencies of war time, when a famous general of æsthetic tastes, in the leisure of a long campaign, employed a common horse-tub for arranging flowers in. Whereas it is a rule with other kinds of flower

FIG. 4.

vases that the water shall not be visible, the water in vessels of the *Sand-bowl* and *Horse-tub* kind is intended to form part of the composition. The idea to be conveyed is that of a broad surface of water or marshy ground, and the plants and trees used are only those which are associated with the propinquity of water. Water plants are chiefly employed, but sometimes plants or trees which grow on the banks of streams are allowed. The [illegible] of such water vessels is, moreover, chiefly confined to the warmer months of the year when the sight of a broad surface of water is grateful and refreshing to the eye. A different rule of proportion between the flower composition and the vessel is followed when these broad shallow receptacles are employed. The height of the floral composition is made about one and a half times the *breadth* of the vessel, and is not regulated by its *height* as with the ordinary standing vases.

It is customary with the above-named water-bowls to use fancy fasteners of metal, to hold, or give the appearance of holding, the base of the flower stems. These will be described afterwards when the whole question of fasteners is discussed. Examples of *Sand-bowls* and *Horse-tubs* are shown in Plate XXVII A. A special water vessel requiring mention is that which goes by the name of the *Long-boat*, not from any particular resemblance in form to a boat, but from the fact that it is very long in proportion to its width, and is made of white wood. In shape, the *Long-boat* resembles a *Sand-bowl*, and it is used in a similar manner, with sand, pebbles, rocks, and water.

Fig. 5

Another form of standing vessel sometimes used for plants is the *Well-frame*, a sort of box-shaped vessel of worm-eaten wood, with a capping piece crossing at the angles like an Oxford-frame, the whole resembling the wooden framing round a Japanese well (see Fig. 5). The rustic character of this kind of vessel requires that the flowers placed therein should be of the simplest kind. In the example illustrated, a plant called Togiri (*Clerodendron squamatum*) is used in combination with a long slender grass called Susuki (*Eularia japonica*).

FLOWER BASKETS.

The famous Regent Yoshimasa is said to have been the first to employ the plaited basket as a receptacle for flowers. Baskets, made of reeds, stems of creeping plants, cane, or bamboo, of Chinese manufacture, were much prized in Japan, and high prices were given for antique specimens. To the present day the more mellowed with age these vessels appear, the more they are valued. A Chinese artizan, named Hokoji, is said to have introduced their manufacture into this country. He presented one of his own make to the retainers of Yoshimasa, with a humble request that it should be embellished with an ornamental stand when placed before the Regent. Yoshimasa is said to have been so pleased with its simple elegance that he ordered it to be put at once on the dais without any stand or tray. From this it is said arose the custom of dispensing with a tray under *Flower-baskets*, though used under other flower vessels. A special kind of *Flower-basket*, with a large oval handle arching over the top, is still called the *Hokoji* shape, from the name of its first maker; and another kind, with a straight neck and no handle, is called the *Reishojo* shape from the name of his daughter, Reishojo. The *Hokoji*-shape basket is shown in Fig. 6, arranged with a convolvulus twined round the handle. Flowers placed in such baskets are fixed in bamboo tubes containing the water, which are hidden within.

Fig. 6.

Another popular story, in connection with the use of baskets as flower vessels, is that the famous philosopher Rikiu started the fashion on the occasion of a visit to the cherry blossoms on the banks of the river Katsura, near Kioto, when he made use of a common trout-basket to arrange gathered flowers in. There is a special basket still in use called either the *Katsura-basket* or the *Trout-basket*. Besides the orthodox shapes alluded to, there are various fancy forms of baskets employed in both standing and hanging positions. Of these the following principal examples may be mentioned:—

The *Half-plaited-basket*, a cylindrical basket in which the vertical threads project two or three inches above the horizontal plaiting, as if unfinished in making.

The *Rain-coat-basket*, a hemispherical basket with a frayed top hanging over the edge of the mouth, and somewhat resembling the straw collar of a Japanese farmer's rain-coat.

VARIOUS SORTS OF FLOWER BASKETS

The *Wool-basket*, a rudely shaped cylindrical basket [illegible] such [illegible] is used in collecting the wool of the cotton plant.

The *Bait-basket*, a small plain basket with a handle, [illegible] that [illegible] for carrying fishing-bait.

The *Horned-basket*, a basket of tall cylindrical form, with two horn-like projections on the top as handles.

The *Square-basket*, a basket of simple cubical shape with no handle.

The *Long-basket*, a tall, thin, tube-like basket.

The *Sosen-basket*, a basket of pyramidal shape, having four sloping sides gathered to a circle at the mouth, and with a high arched handle. The name Sosen is that of its inventor. It is shown arranged with chrysanthemums in Fig. 7.

Fig. 7

Among special baskets for hanging or hooking may be mentioned the following:—

The *Horn-shaped-basket*, a basket shaped like a bull's horn.

The *Cicada-shaped-basket*, so called from its resemblance to the body of an insect.

The *Butterfly-shaped-basket*, a basket which has a short cylindrical neck, and a broad arched body curving out to points at the bottom, the whole form being suggestive of a butterfly with open wings.

The *Hood-shaped-basket*, a basket of an irregular oval shape below with a top opening towards one end, somewhat resembling in shape a baby's sock and, to the

Japa... ... the ...ds worn by woman in the cold season. It is illustrated by Fig. arranged with Willow and Narcissus.

The *Gourd-shaped basket*, a basket woven in the form of a gourd.

The *Daikoku's bag basket*, a basket resembling in shape the bag carried by the Japanese God of Riches, *Daikoku*.

The *Arima-basket*, a basket of cylindrical form with a side mouth, similar to the hanging bamboo basket. The name is derived from a village called Arima.

For certain suspended arrangements of creeping plants a deep net-work basket, something like an English waste-paper basket in shape is often used. Within this a low flower vase is placed from which the creeper rises, being arranged so that part is seen through the netting of the basket, and part hangs over the side in a *streamer* or trail. In all cases flower baskets require an inner vessel for holding the water in which the flower stems are placed, and this generally consists of a concealed tube of bamboo. Examples of various flower baskets, including those above described, are to be seen in Plate XXVII. and Plate XXVIII.

Fig. 8.

BAMBOO VASES.

To Yoshimasa's patronage is also attributed the original use of flower vases formed out of bamboo tubes. As first introduced, these were simple cylinders of thick bamboo, cut near the root, about a foot or more in height, and four or five inches in diameter, the bottom being closed by a natural division. The facility with which such vessels could be cut into different shapes led to the invention of a variety of forms, each bearing a fancy name and specially adapted to different styles of flower arrangement. Portions of the sides were notched out, and side apertures were introduced, often in several stages, so as to allow of two, three, or five compositions in one vase. The different kinds employed are so numerous as to require tabulation, and many of the names almost defy translation; they are therefore given in Japanese. The meanings of these names are explained as far as possible, and it will be seen that they refer chiefly to some fancied resemblance in general shape, or in the form or number of the apertures, to other native objects. They are as follow:—

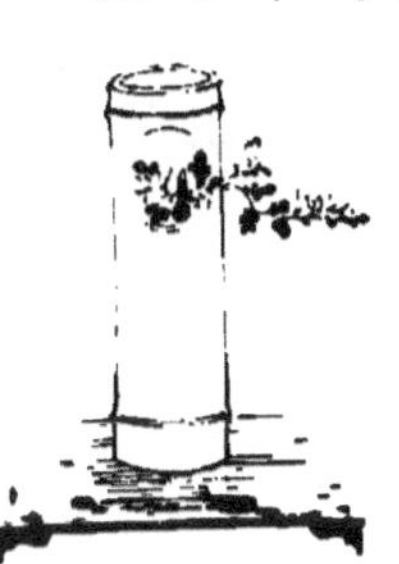

Fig. 9.

Shishiguchi-gata.—Lion's-mouth-shape: a cylinder from ten to fifteen inches in height with a square side-mouth about 3 inches deep.

Noborijishi-gata.—Rampant-lion-shape, a higher cylinder with side mouth as above.

Tabimakura-gata.—Travelling-pillow-shape; a very short vase with small side slit.

Utaiguchi-gata.—Singing-mouth-shape: so named from a splayed form given to the side opening, suggesting the incline of the lips in singing.

Waniguchi-gata.—Shark's-mouth-shape: so called from teeth-like prominences given to the bottom of the side aperture.

Fukurokuju-gata.—Named after a Japanese magician represented with a very high cranium, on account of the low position of the side mouth.

Karamon-gata.—Chinese-gateway-shape: a high vase with one of the side openings rounded, resembling a Chinese arch.

Gammon-gata.—Wild-geese's-gateway-shape; a vase with curved side-opening somewhat like the above, the name being taken from that of a famous arched sea rock called the *Wild-geese's-gateway.*

Kawataro-gata.—Named after a fabulous animal called *Kawataro* or *Kappa*, something like a monkey, and said to inhabit lakes.

Seiro-gata.—Distilling-vessel-shape; a high vase with two small side openings one above the other.

Kawara-gata.—Tile-shape: so called from its resemblance to a half cylindrical capping tile.

Ro-gata.—Oar-blade-shape; resembling the blade of a Japanese oar.

Toro-gata.—Lantern-shape; supposed to resemble a stone standard-lantern.

Auko-gata.—Ray-fish-shape.

Daruma-gata.—Seated-hermit-shape; this vase spreads out at the bottom and has an arched side-opening near the top, suggesting the form of a hermit's cowl.

Noborizaru-gata.—Climbing-monkey-shape; this is a high vase with a very long deep cut in the middle, leaving two short cylinders at top and bottom.

Daibutsu-gata.—Seated-Buddha-shape; so called from its supposed likeness to a seated Buddhist statue.

Enko-gata.—Monkey-shape.

Katoguchi-gata.—Cusped-opening-shape; so named because of the cusped form of the side-opening.

Tora-gata.—Mantis-shape.

Gojiu-gata.—Five-storey-shape; a very high vase with five side openings.

Hsokei-gata.—Icicle-shape; so called from the leg-like cuttings at the bottom of the vase in the shape of icicles.

Mitsuashi-gata.—Three-legged-shape: in this vase the bottom of the cylinder is cut away leaving three legs remaining.

Torikago-gata.—Bird-cage-shape.

Teoke-gata.—Hand-pail-shape; resembling a Japanese hand pail, two deep apertures being cut exactly on opposite sides so as to leave a handle-like strip above.

Tegine-gata.—Pestle-shape.

Usu-gata.—Mortar-shape.

Shakuhachi-gata.—Flute-shape; a long thin tube of bamboo, slightly bent like a native flute.

Hashigui-gata.—Bridge-post-shape; supposed to resemble the newel of a wooden bridge rail, having a deep square slit in the middle.

Miotsukushi-gata.—Beacon-light-shape.

Nijiu-giri-gata.—Two-storey-shape; a vase with two side openings one above the other, in addition to the top opening.

Sanjiu-giri-gata.—Three-storey shape.

Tsurube-gata.—Bucket-shape; named after its supposed resemblance to a well-bucket.

Tsurukubi-gata.—Crane's-neck-shape; so called from the length and depth of the side cutting which leaves a long thin neck of bamboo suggestive of a crane's neck.

Tsurigane-gata.—Bell-shape.

Koma-gata.—Spinning-top-shape.

Tarai-gata.—Tub-shape.

Horagai-gata.—Conch-shell-shape.

Taki-nobori-rio-gata.—Cascade-ascending-dragon-shape; a high bamboo vase cut into a long spiral, supposed to resemble a writhing dragon.

Eboshi-gata.—Ceremonial-cap-shape; named after its resemblance to a Japanese Court cap.

Jikiro-gata.—Food-box shape.

Nijiu-yagura-gata.—Two-storey-castle-turret-shape; so called from square embrasure-like side openings near the top.

Hatomune-gata.—Pigeon-breasted-shape; so called from a bend in the bamboo cylinder giving it a pigeon-breasted appearance.

Rikkwan-gata.—Pan-pipes-shape; a row of small bamboo tubes of different heights tied together with cord and fixed on a stand.

The invention of most of these bamboo vases is attributed to different professors of the Tea Ceremonial. As will be perceived in the above list, assisted by the illustrations, the breadth, depth, and roundness or squareness of the side apertures, as well as their number, and the total height of the vase, suggest the names for the different vases. Many of them are provided with a circular nail hole on one side, near the top, for hanging purposes, and such vessels can be used at option, either hooked to a nail or standing upon the floor of the alcove. The tall kinds having open tops are invariably used standing.

Another variety of bamboo vase not previously mentioned consists of three or more bamboo cylinders of different heights attached in a line, and named *The Row-of-piles*, after their resemblance to a row of pile heads. Many of the above mentioned bamboo vases are illustrated in Plates XXIX. and XXX.

There also exists what is called the *Verdant bamboo vase*, being a vase of one of the above shapes, freshly cut from a growing bamboo stem, with twigs of green leaves remaining on it. In such a vase the intermediary knots or divisions are left intact, and small apertures are introduced in the side for filling in water and other preservatives against speedy withering.

Japanese flower vessels may be broadly divided into three kinds, those used for standing upon a dais, table, or shelf; those intended for hooking against the wall or against a pillar; and those suspended by chains or cords from a ceiling or beam. The vessels hitherto described belong to the *standing* kind, with the exception of some of the baskets and certain of the bamboo vases just enumerated, which can be used either for

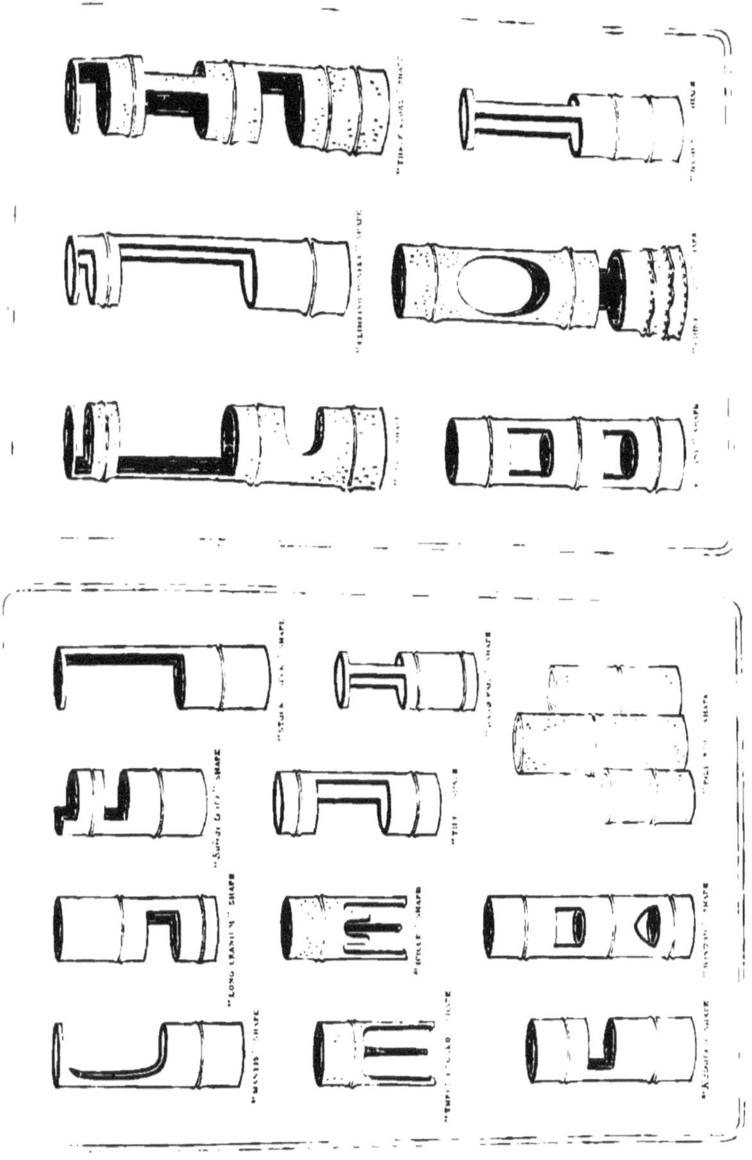

PLATE XXX.

A

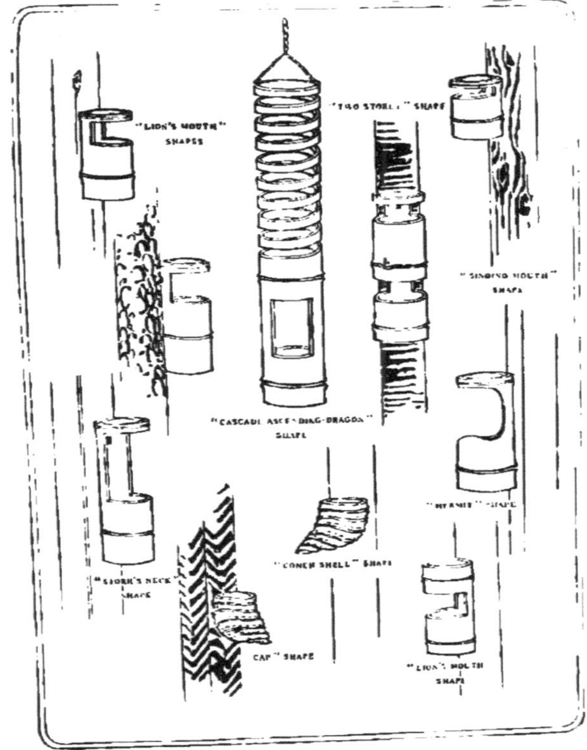

HOOKING BAMBOO VASES

B

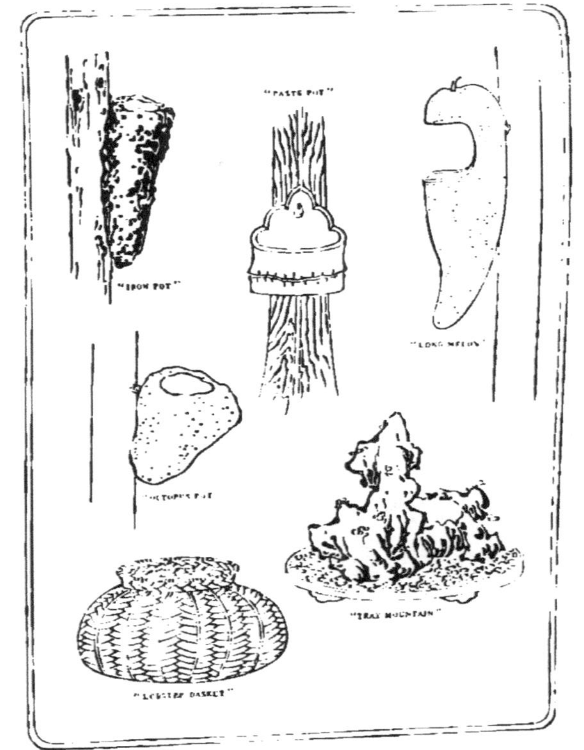

MISCELLANEOUS FLOWER RECEPTACLES

PLATE XXXI

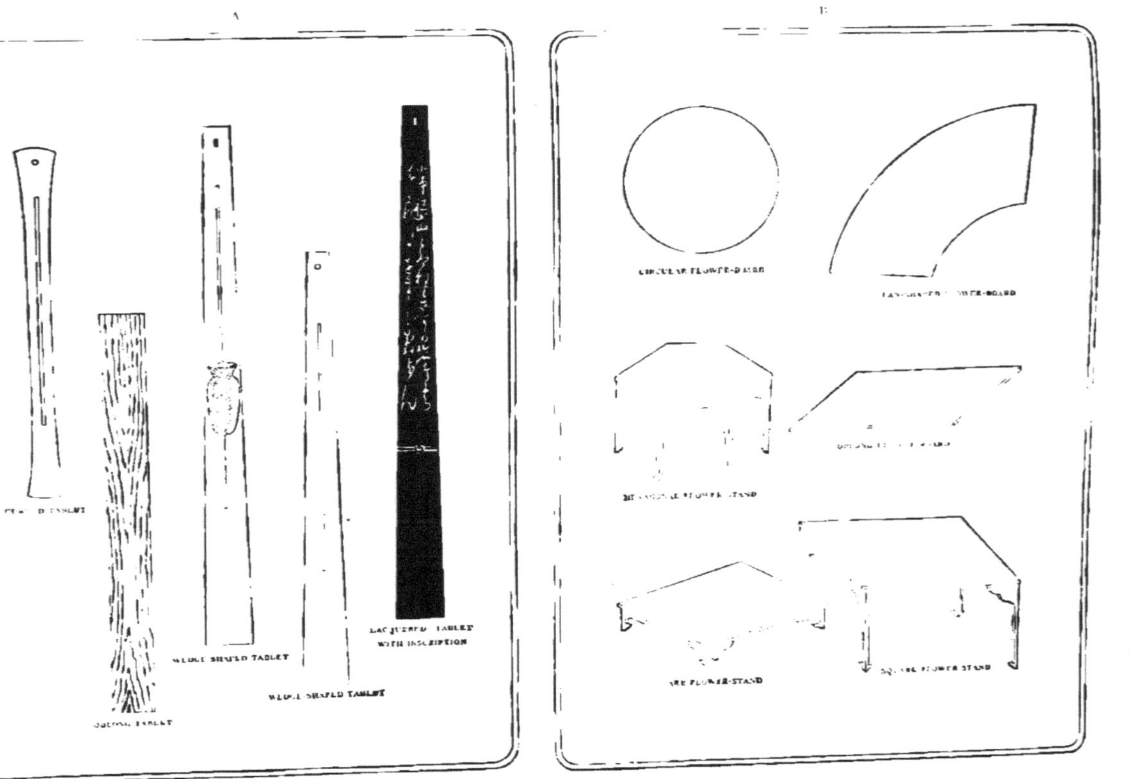
A
B
WEDGE SHAPED TABLET
WEDGE SHAPED TABLET
LACQUERED TABLET
WITH INSCRIPTION
CIRCULAR FLOWER-BOARD
ARC FLOWER-STAND
SQUARE FLOWER STAND

standing or hanging. In order to distinguish clearly between flower vases which are hooked to a pillar or wall surface and those which are hung by chains or cords, the former will be called in future *Hooked-vessels* and the latter *Suspended-vessels.*

HOOKED VESSELS.

Hooked-vessels are of various kinds, from the chrysalis-shaped root of a bamboo, to the form of a shell, gourd, or melon. They are invariably short, compared with the standing vases, and with a few exceptions, the absence of flatness below suggests their method of use. Among the bamboo vases, those of little height and with narrow side apertures, such as the *Lion's-mouth-shape* and *Travelling-pillow-shape,* are used mostly as hooked vases. A lateral direction is given to floral compositions placed in hooked vessels, the idea suggested being that of flowers hanging over a cliff. For tea rooms, where a severe and rustic style of flower composition is preferred, vases of curious shape are pressed into use, among which may be mentioned :—the *Octopus-pot,* a coarse irregular shaped earthenware jar used by fishermen for holding the octopus ; the *Iron-pot,* a rough iron pot-shaped vessel somewhat like a martin's nest, and the *Decayed-stump,* a piece of decayed wood hollowed out as a vase. The *Gourd* is also a favourite form for hooked vases, the mouth being sometimes cut in the side and sometimes at the top. It is said to have been first used for flower arrangements by the philosopher Rikiu, who once extemporized a floral design in a wine-gourd which he took from an itinerant priest at the temple of Sumiyoshi near Osaka.

Fig. 20

As previously mentioned, many of the woven baskets employed as flower vessels belong to the hooked class. These are to be found described under the head of flower-baskets, and are illustrated in Plate XXX.

As a background to the *Hooked-vessels,* and originally intended to protect the pillar or prepared wall-surface from staining or abrasion, narrow oblong tablets of wood are often used. They are sometimes made ornamental, being lacquered and inscribed with verses in gold letters. Some are plain oblong tablets about four inches broad and three or four feet long; others are wedge-shaped, tapering towards the top; and others have curved sides. They are provided with a long narrow slit down the middle for sliding to different heights over the iron nail or peg by which they are held to the wall or pillar, and to which the flower vase is hooked. In some cases these tablets are hinged in the middle to allow of folding up when out of use. They are often made of segments of bamboo flattened out and polished or lacquered. Examples of these hanging tablets may be seen in Plate XXXI.

SUSPENDED VESSELS.

Suspended vessels are those hung by a cord or chain to the ceiling or lintel of a recess. Belonging to this class is a crescent-shaped vase of pottery or bronze called the *Crescent-moon*. The horns of the crescent are made almost to meet and are suspended from above by a connecting ring and single chain. The other kinds being of more elongated form are hung from both ends by double chains or cords. The simplest of these are bamboo tubes splayed off at the ends, hollowed out in the middle, and hung horizontally, so as to suggest the form of a boat or punt; others are of bronze, shaped in exact resemblance to a ship or junk. Yoshimasa is said to have conceived the idea of boat-shaped vases whilst observing children sailing toy boats filled with flowers. Another story attributes the first use of such vessels to the famous philosopher Soami, who on a hot summer day, to please his patron Yoshimasa, took a bronze vessel of accidental resemblance to a boat, and by his manner of arranging the stems of the flowers therein, conveyed the

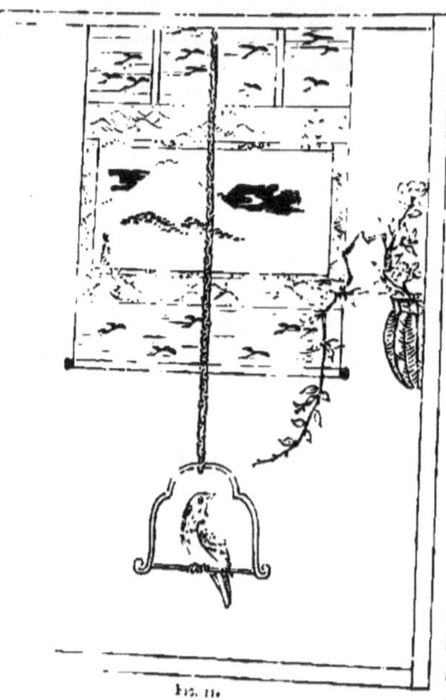

Fig. 11.

PLATE XXXII.

A

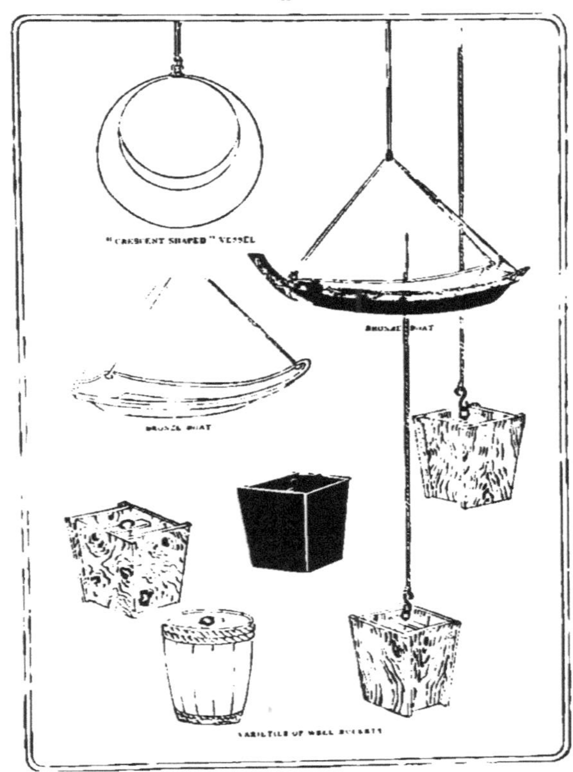

B

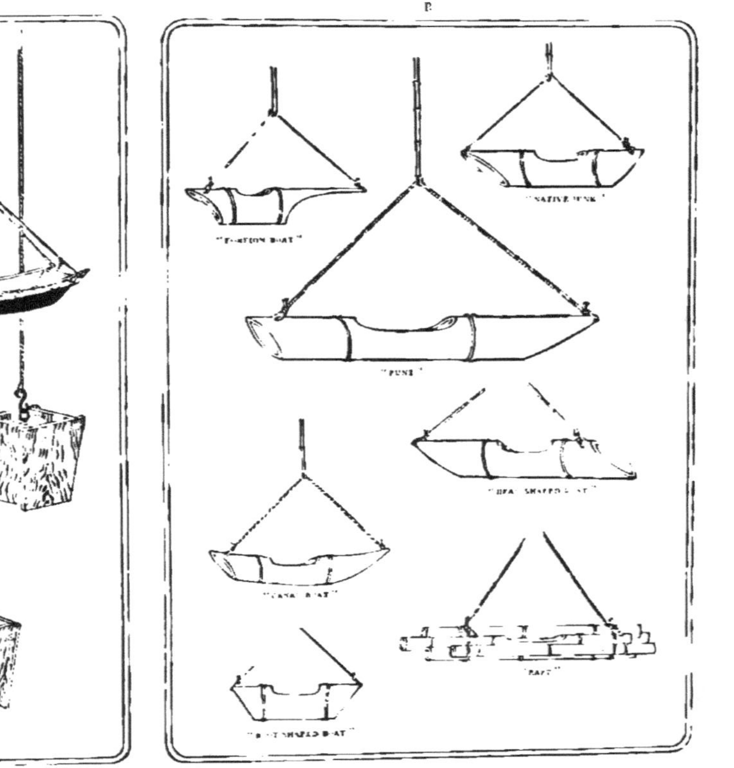

an anchor of black metal hanging to the side: the *Basket-work-boat*, a boat of metal and basket-work. Also a curved cane-work tray, oblong in shape, hung from the ends, and carrying in the middle a little bowl of flowers, is sometimes used.

An important theory in boat arrangements is, that they ought always to be suspended in an elevated position, both with a view to preserve the idea of a floating vessel, and also to prevent the possibility of seeing the water which they contain. It is held to be a great violation of taste to allow the water which is necessary for preserving the plants in a flower-boat to be seen, because, water visible within a ship would be suggestive of a leaking or wrecked vessel, and would be consequently considered unlucky. Sometimes the idea of a stranded or beached boat is purposely conveyed by a flower vessel which is placed upon the dais instead of being suspended. In this case the vessel should be raised upon a stand of some kind so as to place its upper surface above the eye level of seated visitors. Such standing boat-vases are supported upon two wooden rollers or upon a light frame of cross-pieces. Before leaving the subject of boat-shapes, allusion must be made to a standing vessel called the *Long-boat* which is sometimes employed for elaborate arrangements of plants and grasses. This vessel appears to be called a boat simply on account of its narrow length and the fact that it is made of plain white wood. It resembles a *Sand-bowl*, being an oblong tray-like vessel with short legs. It is five feet long and about one foot wide and is only used for very large recesses on special occasions. The boat-vases described above are illustrated in Plates XXXII. and XXXIII.

Fig. 13

The classification given refers only to the different shapes and sizes of the boat-like vases. Other terms are used to designate the manner of hanging the vessels and of arranging the flowers within, so as to convey different nautical ideas. The three principal arrangements are those of the *Outward-bound-ship*, the *Homeward-bound-ship*, and the *Ship-in-port*. Besides these there are other designs known as the *Distant-ship*, the *Swiftly-sailing-ship*, the *Becalmed-ship*, and the *Branch-laden-ship*. These different fancies are conveyed first, by the direction, right or left, and backward or forward, given to the prow of the vessel; and secondly, by the distribution of the different lines of the flower composition. Even the length of the suspending chain and the distance or proximity of

PLATE XXXIII.

These different styles of composition are shown in a skeleton form in Plate XXXIV. Other special rules for hanging boat-vases will be considered afterwards, when the general question of the position of flower arrangements in a chamber is discussed.

Suspended vessels called *Well-buckets* are often used in pairs hung over a pulley by a thick silk cord. One of the buckets is allowed to rest on the floor, or in some cases upon a frame designed in imitation of the railing or boxing round a well, and the other is suspended in the air.

To Rikiu is attributed the first use of such flower vessels, the idea coming to him whilst he was observing a convolvulus twining round the bucket of an old well. A similar pair of buckets are occasionally employed without the pulley and suspending rope, one being placed balanced on the edge of the other so as to leave only a portion of the lower one uncovered for the insertion of flowers. In this case the rope is arranged in a flat coil as a stand for the lower vessel. Buckets used in this way are always flat-sided, to ensure stability, but for the ordinary suspended arrangements, cylindrical as well as square buckets are employed. These vessels are of plain wood, of wood lacquered black, or of worm-eaten or decayed timber. For the most handsome kind in black lacquer, a chain of silver or a red silken cord should be used; to those in ordinary wood a plainer cord may be attached; and in the case of the rustic buckets of decayed wood, a common hemp rope or even an iron chain may be substituted. Single buckets are occasionally to be seen standing upon a low table or decayed slab of wood, or hung by a single bamboo rod. (See Plate XXII.A.)

Fig. 14

Porcelain buckets and pulleys, although not uncommon, are of quite modern introduction, and not according to rule.

Other fancy vessels suspended by cords or chains are sometimes employed. Among the bamboo vases in Plate XXX.A. will be seen one which is suspended by a chain, like a lantern. Suspended baskets, distinct from the hooked baskets previously described, are not uncommon. Another example is the suspended net-work basket previously described.

the arrangement from the observer is governed by the style adopted. According to the principles of lineal distribution which apply to all suspended flower designs, the *Streamer* holds an important place in the above examples. It is in such cases intended to suggest the long bent oar which in Japanese boats trails back towards the stern. This floral line must not be too powerful, as it represents the idea of an oar dragging in the water. The central flower stem stands for the single mast of a junk with or without sails, and the subsidiary stems indicate the other sails and rigging of the vessel. This will be better understood by describing one or two of the arrangements in detail.

Homeward-bound-ship. In arranging a boat of flowers in this form the prow of the vessel is turned towards the left, which, in superior rooms is the host's side of the chamber, in order to convey the idea of home-coming. The central stem of the floral arrangement is high and full, curving towards the helm, so as to indicate a ship in full sail, and a *Streamer* hangs over the front side sloping back towards the stern on the right. The above is a favourite device on occasions of rejoicing for a safe return, or when a son or daughter-in-law is being received into the family. Some say that this method of arrangement should only be employed from noon till dusk.

Outward-bound-ship. This is an arrangement exactly the reverse of the former, the vessel having its prow turned to the guest's side of the chamber, on the right. It is adopted at parting gatherings, in token of wishing good-speed to those setting out on a journey. It is said that this style of composition should be employed only from morning till noon.

Ship-in-port. In this arrangement the vase has the same direction as the *Homeward-bound-ship*, but the floral design is kept small and straight, so as not to suggest wind or motion, and the *Streamer* hangs over the further side of the vessel. Such a disposition of the flowers should not be made excepting during the hours of evening.

Swiftly-sailing-ship. The direction of the vessel in this design is to the right, or outwards; the floral arrangement is full and bent, but no *Streamer* is used.

Branch-laden-ship. The direction of this vessel is inwards, or towards the left, and the floral arrangement is kept short and close, and consists of small flowers, such as daisies or carnations, which are not allowed to project beyond the limits of the vase itself. The idea suggested is that of a ship loaded with timber or tree branches.

with four [illegible] wheels and no shafts, like a child's toy-cart, bearing a small bucket in which the flowers are arranged.

FLOWER FASTENERS.

The subject of *Fasteners* for floral arrangements is one belonging to the technique of the art under consideration. To a great extent, however, the methods of fastening are treated as a part of the decoration of the compositions, and as such they are closely connected with the different flower vessels employed, and require notice in the present context.

As before mentioned, the *springing*, or point of origin of the floral group, is of great importance, and the firm and skillful fixing of the stems or branches in the vessel which holds them is one of the most difficult parts of the manipulation. Ordinarily, the stems are held in position by small cylindrical pieces of wood fitting tightly across the neck of the flower vase, and having a slit, wider above than below, for threading them through. The wedge-shaped form, wider towards the top, which is given to the slit, allows slightly different inclinations to be imparted to the several branches. The fastener should be fixed about half an inch below the surface of the water, the level of which is made to vary according to the season, and it should not be visible from the front of the vessel. If the vase used be a lacquered one, paper should be placed between its surface and the ends of the fastener to prevent scratching. In some large-mouthed vessels, and in the *Flower-baskets*, the flower stems are fixed in concealed tubes of bamboo which hold the water and the fasteners. Some Schools affect a rustic simplicity in their appliances and employ a naturally forked twig to hold the flowers in position.

For arrangements of water plants in neckless vessels such as *Sand-bowls* or shallow *Tubs*, other sorts of *Fasteners* are necessary, which are hidden below the sand or pebbles which such vessels contain. One kind consists of a sheet of copper perforated with holes of different sizes to receive the extremities of the different stems. Another *Fastener* is made of rings or sections of bamboo of varying diameters attached to a wooden board, the stems finding lodgment in the sockets thus formed, and being further held in position by the pebbles which cover them. Occasionally a *Fastener* called the *Whirlpool*, and consisting of a spiral hoop of metal placed vertically, is employed.

PLATE XXXIV.

SUSPENDED BOAT-SHAPED BAMBOO VESSELS, SHEWING LINES OF FLORAL ARRANGEMENT

as the dragon is a mythical monster belonging to all elements, the use of this fastener is not limited to any particular kind of plant.

The *Tortoise-fastener* consists of one or two tortoises in bronze arranged in different positions.

The *Water-fowl-fastener* is generally a metal imitation of a pair of mandarin ducks. It is occasionally attached to water plants.

The *Frog-fastener* needs no special explanation, except that, representing an amphibious animal, it may be used with both land and water plants.

The *Anchor-fastener* is specially intended for use in suspended boat-shaped vessels. It is incorrect to fix it in a vase representing a stationary ship, as in such a case the anchor would not be visible.

The *Knife-fastener* is a metal knife or dirk such as is worn in the wooden sheath of a Japanese sword, and owes its original use as a flower fastener to a floral arrangement once hastily extemporized by a famous artist named Oribe, in which, having no other fastener at hand, he used his knife for the purpose. The *Scissors-fastener*; the *Pipe-fastener*,—a long metal tube with a small bowl; the *Weight-fastener*,—an oblong metal paperweight, the *Chain-fastener*,—a short chain disposed in a bunch; and the *Kettle-stand-fastener*,—a small iron ring and tripod used for supporting the kettle over the charcoal brazier;—these also are all occasionally employed. The principal of the above fasteners are illustrated in Plates XXXV. and XXXVI.

A special kind of fastener, called by the Japanese *Jakago*, needs separate notice. The native name *Jakago* refers to the long sausage-shaped bags of bamboo basket-work which are filled with bowlders and laid in fascines at the sides of rivers to break the current and protect the banks. They are a common feature in river scenery and have therefore come to be imitated in flower arrangements intended to be suggestive of the presence of water. The *Jakago-fasteners* are long cylindrical baskets with closed and rounded ends which are laid in shallow basins together with ornamental stones or rocks, and besides being decorative they serve to hold the stems of the plants artificially arranged within. An illustration of their use may be seen in Fig. 33, where two are shown combined with the *Kerria japonica*, the whole being intended to represent a view of the river Tama near which these flowers abound.

The inverted bronze bell suspended by a chain is another [illegible] occasionally used for holding arrangements of wistaria flowers. (See Fig. 16.)

A curious form of suspended flower vase is the inver[illegible] umbrella, an exact imitation in bronze of a Japanese umbrella. (See Fig. 14.)

A large sea-shell hung by a single cord forms a favourite receptacle for very simple flower arrangements.

FLOWER CHARIOTS.

Belonging strictly speaking to the class of standing vessels, but sufficiently striking and important to require special notice, is the *Flower-chariot*, which figures so often in pictures on painted screens and other decorative objects. At certain festivals and processions it appears that large tubs full of richly arranged flowers were drawn upon wheeled chariots handsomely ornamented. The idea was adopted for flower arrangements placed in very large recesses, where great size and display were required. The length of the *Flower-chariot* is four feet six inches from the back to the end of the shafts; the wheels are about eighteen inches diameter, and the flower tub which the chariot carries, is about sixteen inches high. Both vehicle and flower tub are lacquered black and furnished with silver fittings. The flower compositions are made very full and high. The *Flower-chariot* is illustrated in Fig. 15.

FIG. 15.

A somewhat similar flower receptacle called the *Water-carrying-cart* is also employed. This vehicle is a flat truck

For arrangements in these shallow vessels there are a number of fancy *Fasteners* in common use which are in many cases merely ornamental, the hidden contrivances just described, buried below the sand or pebbles, doing the real duty of holding the stems in position. The principal of these ornamental fasteners are as follow:

The *Horse's-bit-fastener* is an exact counterpart of a ponderous Japanese bit. Its use originated with the employment of the *Horse-tub* as a flower vessel, and to this kind of vessel its use is chiefly confined. The linked character of this fastener allows of its being folded in a variety of ways, so as to leave loops of different size to encircle the flower stems. With the scrupulous minuteness of detail which characterizes the art under discussion, the floral designer has classified these different ways of using the horse's-bit, giving names to each bar, plate, and loop of iron, and inventing terms for the various methods of folding. The principal arrangements are shewn in Plate XXXV.a. The use of the *Horse's-bit-fastener* is prohibited, however, for floral designs placed in the ornamental recess of a chamber of superior class; and if it be introduced into a flower arrangement in such important rooms, the composition must not occupy the principal position.

The *Crab-fastener* consists of a metal crab or pair of crabs. If one crab be used, it should be disposed so as to contrast in character with the flower arrangement, by which is meant, that if the composition be high and powerful, the crab must be placed in a low and unobtrusive position, but if the flower design be broad and wanting in vertical strength, the crab must be raised in a climbing attitude. If a pair of crabs be used, one must be elevated and the other lowered in position, or, to adopt the quaint phraseology of the floral art, one must be *male* and the other *female*. As the representation is that of a land and not a sea crab, this kind of fastener may be used with land as well as with water plants.

The *Hare-fastener* is a bronze hare in miniature. It may not be affixed to water plants, and is specially suitable for arrangements of wild plants and grasses, such as the lespedeza, rush, and elecampane.

The *Pair-of-carp-fastener* consists of a pair of metal fish designed in the position of two carp sporting together. This fastener is, as might be supposed, only used for water plants.

The *Dragon-fastener* is a metal ornament in the form of a writhing dragon, and,

Serene—expressed by a suspended bronze boat, bearing white chrysanthemums, suggestive of a loaded ship in port.

Austerity.—the sentiment conveyed by the *Eularia japonica* and *Patrinia scabiosæfolia* arranged together in a small bronze vase.

Quaintness.—denoted by a hooked vessel in the shape of a gourd, containing small chrysanthemums.

Brightness.—the idea suggested by a bronze vase engraved with a design of wild geese flying across the full moon, and holding lespedeza flowers.

Chastity.—the character expressed by a bronze vase engraved with a design representing rain, and containing a branch of maple.

Fig. 16.

Security:—denoted by some kind of water plant, placed in a bronze vase engraved with the design of a spider's web.

Veneration:—the sentiment conveyed by a branch of pine, or some other evergreen, placed in a bronze vase engraved with the representation of a crane. The crane and pine tree are both associated in Japan with the idea of venerable old age.

The above combinations, capricious as some of them may appear, serve to illustrate the manner in which vessels and flowers are used together to express an appropriate sentiment.

Sometimes the harmonious connection between the two is based merely upon a resemblance in the name of both. The clematis, for example, is called *Tessen* (*Tetsu-sen*) and because the word *Tetsu* signifies *Iron*, this flower is often placed in a rough iron vessel. (See Plate LIX.B)

The native name for the wistaria is *Fuji*, and the bell-like ornaments hung to the eaves of temples being called

PLATE XXXV

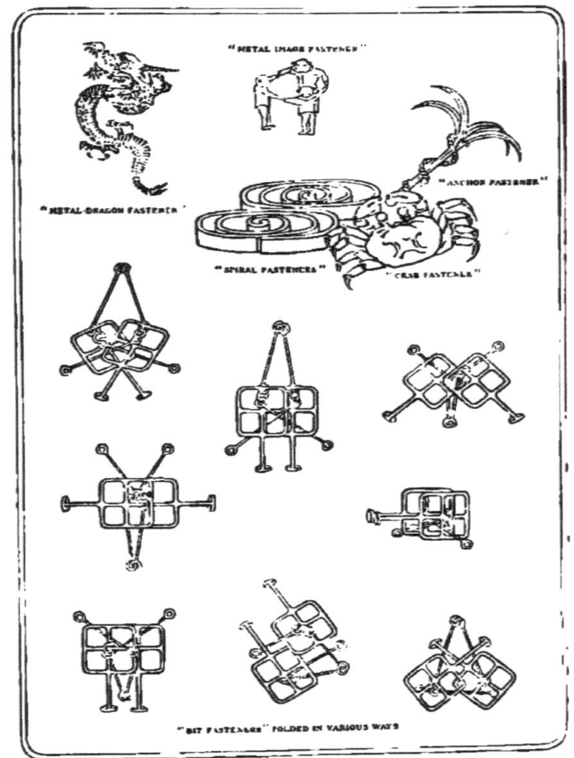
"METAL-DRAGON FASTENER"
"METAL IMAGE FASTENER"
"ANCHOR FASTENER"
"SPIRAL FASTENERS"
"CRAB FASTENER"
"BIT FASTENERS" FOLDED IN VARIOUS WAYS

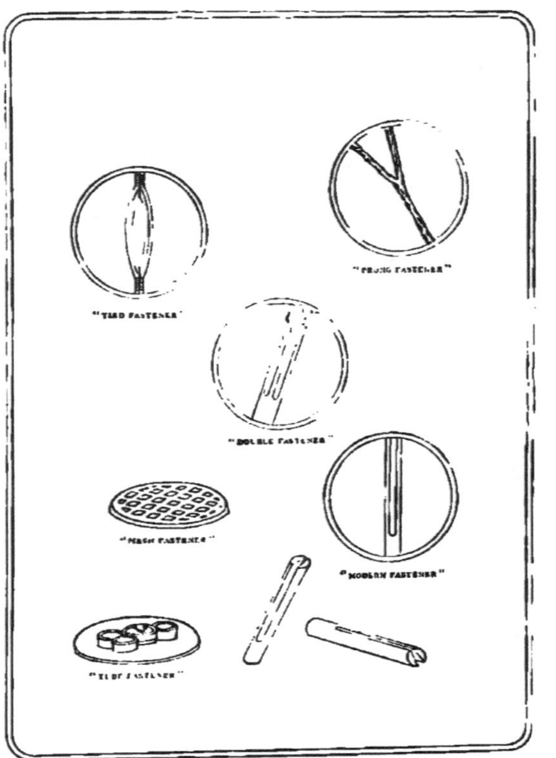
"TIED FASTENER"
"PRONG FASTENER"
"DOUBLE FASTENER"
"MESH FASTENER"
"MODERN FASTENER"
"TUBE FASTENER"

Other general directions are given as to the style of arrangement suited to special kinds of vessels.

In *Hooked vases* the floral design should suggest plants hanging over a cliff, and must be arranged in the horizontal triangle style, with or without a *Streamer*. In vases of this class having a mouth at the side, the lines of the flowers must not cut the edge of the aperture.

In *Standing-vases* of bamboo with two openings, the upper mouth should hold a *tree* and the lower one a *plant*, in accordance with natural scenery in which the tree branches occupy a higher position than the plants.

Fig. 17

Often the same flower is used in both mouths of the same vessel, in which case some such distinction as the following exists:—Supposing pine branches to be arranged in both openings, a style called the *hill and valley pines* is adopted, in which the top branch represents the pine trees on the summit, and the lower branch those at the base of the hill. The idea of distance must be suggested in the former, and that of proximity in the latter.

If the bamboo vase have one top opening and two side apertures, a composition called the *hill, plain, and water* style is followed. To convey this idea a mountain *tree* is placed in the top, a land *plant* in the middle, and a water *plant* in the bottom opening.

Sometimes an arrangement exactly the reverse of this is devised in order to express the notion of a distant landscape with a mountain lake above, fields on the hill slopes, and a forest at the base. In such a composition, the uppermost mouth of the vessel contains a water *plant*, a land *plant* is placed in the middle aperture, and a *tree* branch occupies the lowest position. The *tree* branch should be arranged high to preserve the idea of a near foreground; the land *plant*, expressing middle distance, may be of moderate proportions; and the water *plant* at the top must be kept small and cramped in character, to suggest distance.

PLATE XXXVI.

A

Kerria japonica IN "HORSE TUB," WITH "HORSE BIT FASTENER"

IRIS IN "WELL FRAME" VESSEL WITH "CRAB FASTENER"

Rosa rugosa IN BASIN WITH "PAPER-WEIGHT FASTENER"

B

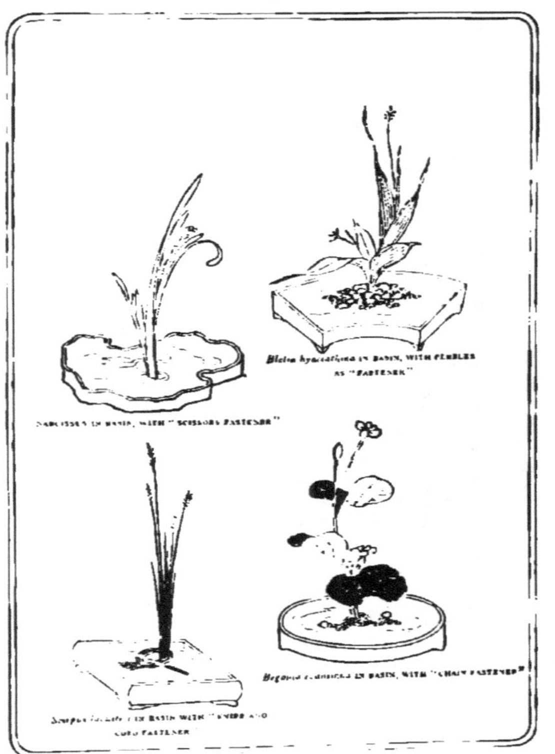

NARCISSUS IN BASIN, WITH "SCISSORS FASTENER"

Bletia hyacinthina IN BASIN, WITH PEBBLES AS "FASTENER"

Scirpus lacustris IN BASIN WITH "KNIFE AND CORD FASTENER"

Begonia evansiana IN BASIN, WITH "CHAIN FASTENER"

In *Sand-basins*, if a *tree* be used, it must be supported by a *plant* of some kind. Plants alone may be combined, but the composition must in such a case be full and strong to give it an appearance of solidity. Trees or plants arranged in these vessels are often divided into separate groups with a space between. (See Plates XXXIII. and XLIX.). If trees are used, the interval between each clump is technically called the *Valley-space*, and it should be equal to about one-tenth of the height of the arrangement. If water plants are used in divided groups, the distance between them is regulated in the same proportion, but different terms are used to denote this spacing. Supposing such water plants to be arranged side by side and in a line, the composition is said to be in the *Fish-swimming* style, but if the flowers are placed one in front of the other they are said to be composed in the *Fish-sporting* style. By a curious fancy an analogy is here drawn between the relative position of the plants in such broad vessels and that of fish swimming or sporting in a lake or stream. The arrangements of trees or plants in water basins, whether in single or double groups, must be always more towards one side of the vessel than the other, all formal and symmetrical compositions being disliked.

In *Horse-tubs* the employment of *tree* branches is prohibited, and *plants* of one or two kinds must be used. With regard to spacing, the same rules apply as are given for *Sand-basins*.

If a pair of similar vessels containing floral compositions be used side by side, the flower arrangement in one should be as nearly as possible of the same form as the other, reversed; but the colours of the flowers should be varied. For example, one vase may contain red and the other white flowers, with the stems or branches of both disposed in nearly similar lines. These symmetical arrangements are not, however, often resorted to.

In *Flower chariots* it is usual to arrange the seven flowers of Autumn; with these are sometimes combined other Autumn grasses, making the number up to nine or eleven different kinds of plants. Such elaborate combinations, which are not generally allowed in other flower vessels, are from their richness technically called *embroidery*.

WATER IN VASES.

Various rules are observed as to the use of water in flower vessels. In Spring and Autumn the vase should be about nine-tenths filled with water; in early Summer

CHOICE OF FLOWER VESSELS.

The Japanese flower artist recognizes a distinct and important connection between the floral composition and the receptacle in which it is arranged. Some points of relationship, such as that of the proportion in height or breadth between the two, and the distinction between vessels used for land plants and those suitable for water plants, have been already mentioned. But there are other more subtle harmonies in taste and sentiment which are carefully observed. A flower vessel, being in itself a work of art, may possess different æsthetic characteristics, such, for example, as rustic simplicity, elegance, or richness. Its shape, material, or decoration may also convey to the imagination various feelings and mental associations. Flowers, too, possess different qualities in form, colour, or growth, and are in all countries associated with special poetic attributes. Particularly is this the case in Japan where hardly a fête or pastime exists in which flowers do not play a part, and where almost every blossoming tree has some romantic tradition of its own. Apart from the character intrinsically belonging to the flowers themselves, their artificial treatment in the flower art under consideration imparts to them other characteristics varying with the style of arrangement adopted. Thus one composition may be full and luxurious, whilst another design made with similar flowers may be simple and even austere.

Considering, then, that both flower arrangements and flower vessels are works of art capable of distinct artistic expression, it becomes a matter of importance that the spirit of the one should accord with that of the other. This harmony of sentiment need not necessarily be one of complete unison; on the contrary, it is often produced by a well-judged contrast.

By way of illustrating this intimate connection between floral compositions and the vessels containing them, may be given the following ten artistic virtues attributed to certain special combinations :—

Simplicity :—expressed by rushes and irises in a two-storey bamboo vase.

Aspiration :—denoted by a vessel of decayed wood containing a climbing creeper.

Affection : the character attributed to a bronze basin containing a pine branch entwined by a wistaria.

STONES AND ROCKS.

In addition to the flowers, vessels, and fasteners, all of which contribute to the completion of a Japanese floral design, stones or rocks are sometimes added to arrangements of water plants and play an important part in the compositions. These stones represent, in some cases, the large bowlders which form stepping stones over streams and lakelets, in other instances, they are meant to suggest islands in extensive water scenery. Again, it sometimes happens that land and water plants are used in combination in a large shallow vessel, and then the stones are disposed so as to suggest the dry bed or the banks of an adjacent river. Both white and black stones are employed, the white ones being placed near to the flowers, and the dark ones in parts of the water where there are no flowers. The flowers are 'arranged in front of or behind the stones and must not appear to grow out of them.

The chief ornamental stones in a flower basin are generally distributed according to the favourite triple principle which is applied also to the lines of the flower composition, under the distinguishing designation of,—*heaven, earth, mankind*, conveying in the present case the idea of verticality, horizontality, and intermediate form. One stone is of vertical character and supposed to resemble a mountain, one of flat and nearly horizontal character, and the third, which is placed between the other two, partakes of an intermediary chracter. Other stones of secondary importance are added to set off the larger stones and generally to connect the whole composition.

The use of such stones may be seen in Fig. 25, where they are arranged in a large basin together with wistaria flowers. In Fig. 18, on the previous page, is shown a fancy arrangement in which blocks of charcoal are employed instead of stones.

FLOWER TRAYS AND STANDS.

All standing flower vessels, with the exception of the *Flower-baskets*, are placed upon a square tablet of polished or lacquered wood, interposed as a protection between them and the surface of the dais or shelf on which they stand. A story is related elsewhere which accounts for the absence of such tablets under flower-baskets, but a very reasonable explanation seems to be that these baskets are not likely to stain or

Furin, by way partly of a play upon the two words, the wistaria is sometimes arranged in a inverted bronze bell. (See Fig. 16).

Certain writers go so far as to classify flower vessels according to the seasons, recommending, for Spring arrangements, bamboo vases, bronze vases, and narrow necked vessels; for Summer compositions, flower-baskets, bronze basins, wooden tubs, or other broad-mouthed vessels; for Autumn designs, boat-vases, and porcelain vessels, and for those of Winter, gourd-shaped vases, and narrow necked vessels. However rare and valuable a receptacle may be, it must not be used for holding flowers unless intended for that purpose, jars, jugs, pots, and other utensils having special uses of their own should not be employed for floral arrangements. This rule is apparently violated in the case of such important flower-vessels as the *Horse-tub*, *Well-bucket*, and *Flower-boat*. But these are exceptions which custom has sanctioned, and their names have reference rather to the original models from which they are copied. In each case, moreover, there is a special connection in idea between these receptacles and the flowers placed in them, so that the result has no element of incongruity.

In such matters, however, considerable license is allowed to masters proficient in the art, especially in the case of floral designs for tea rooms, where the employment of curious vessels of all kinds is permitted. Some of these, such as the *Paste-pot*, *Octopus-pot*, and others, are illustrated in Plate XXX.II.

A few general directions are laid down as to the particular kind of receptacles suited to certain flowers.

For flowers of large-blossom, such as the peony, the Chinese basket is preferred, the peony being considered the principal flower of China. Moreover, these large baskets are in character well suited to show off the ponderous blossoms of this plant.

For most water plants, low basin-like vessels, or vases with very broad mouths, are best suited, but the narcissus requires a narrow necked vessel to show it off to advantage.

For plants of short and stunted growth, having large leaves, a tub-shaped vessel is chosen; and for the wistaria, lespedeza, and *Kerria japonica* some kind of suspended receptacle is preferred.

In arranging flowers in tall bamboo vases which have several mouths, the composition in the upper openings should assume the from of the horizontal or leaning triangle, whilst that in the lowest mouth should be treated in the style of the vertical triangle. (See Fig. 17).

In the case of a pair of *Well-buckets*, the upper vessel should have a *tree* and the lower one a *plant*. The rope should be kept sprinkled with spray as if covered with dew. Neither of the compositions in the two *Well-buckets* must be allowed to cross the line of the rope or chain which suspends them. Instead of the ordinary flower fasteners, *Well-buckets* generally have a perforated lid or frame which is fitted in the top just above the surface of the water. When these vessels are arranged in combination with a *Well-frame*, the upper bucket is suspended and the lower one rests on the edge of the frame, and in such a case the lower vessel may show water, but the upper one must preserve the idea of an empty bucket, and by no means must the water it holds be visible. In Spring time the floral design in the upper bucket should be the fuller of the two, but in Summer time the lower one should contain the more crowded arrangement. In Autumn both compositions should be simple and quiet. For arrangement in double *Well-buckets* the lower floral design should be of the style used for standing vases, and the upper one of the style employed for suspended vessels, with a lateral lean and a *Streamer*. Flowers arranged in square buckets should never be placed exactly in the angles or corners of such vessels.

In the kind of double bamboo vase called the *Row of piles*, the higher tube should contain a land *plant* and the lower tube a water *plant*.

In the *Bridge-newel-vase*, a cylindrical vessel with a top mouth and an oblong side opening, the top of the vase should have a thick stump or heavy arrangement of *tree* branches, and the side aperture should contain some simple *plant*, modestly arranged.

In *Flower-baskets*, those with arched handles should have the flowers arranged so as to keep within the enclosure of the handle and not cross it. In very elaborate compositions this rule is sometimes violated, but in such cases the cutting or crossing must only take place on one side, and by no means in the centre of this side. The handleless baskets are generally hooked vessels, and flower arrangements in them are disposed as for the latter class of receptacles.

The back wall of the *Toko-no-ma* is the surface upon which the *Kakemono* or rolled pictures are exhibited. These paintings are hung singly, in pairs, in triplets, and occasionally in quartetts. The floral design is placed upon the dais of the recess, below and to the front of the pictures, unless a hanging arrangement of flowers be used, in which case it is suspended from the lintel or ceiling, or hooked to the pillar of the alcove.

It is considered important that the floral composition should not in any way clash with the pictorial arrangement, either as regards position, line, subject, or sentiment. The two together must form a harmonions decorative composition.

When only one wall-painting is exhibited, the vase of flowers in front should be placed, as a general rule, rather to one side, if two pictures are hung, a single floral design should occupy the interval between them, in the case of three pictures, two vases of flowers should be used, one opposite to each interval; and with four paintings, three flower compositions are employed in a similar manner. In the last case, instead of using three floral arrangements, the central space is sometimes occupied by a statuette or incense-burner.

Vases containing flowers are often elevated upon a small raised stand or table, employed instead of the flat board or tablet which is placed under most flower vessels. One kind of table used has a shelf below, on which a very simple floral design may be arranged, and when so employed the top of the table supports some other ornament.

The proportions of the *Kakemono* or hanging picture influence the disposition of the flowers on the dais. In front of a long painting, the floral composition must be kept as low as possible, but when a short and broad picture, called a *Yokomono*, is displayed, the flowers may stand high and full in arragement. The object of this rule is to prevent the mural painting from being hidden by the floral design. The same result is often obtained by placing the vase of flowers to one side of the recess, instead of in the centre. It is sometimes unavoidable that the flowers cover part of the picture, but under these circumstances special care must be taken not to hide that portion bearing the stamp and signature of the artist. The centre, ends, and tassels of the ornamental roller forming the bottom border of the painting must never be obstructed. When the pictorial work contains figures, the features of these figures must on no account be hidden by the branches of the flower arrangement, and if, as occasionally happens, the picture is inscribed with a poem or proverb, this writing must be exposed to view.

nearly full, and during the hottest days of Summer brim-full to overflowing, the rim of the vessel being oiled so that the water actually seems to pile up above the edge. In Winter time the vase should only be four-fifths full, and as little as seven-tenths in the coldest season.

When water plants and grasses are arranged in broad shallow vessels, the water forms part of the composition, and the different portions of its surface are accordingly regarded as having different properties. The water nearest the flowers is supposed to be moving and life-giving, and must therefore be free from floating matter; the portion removed from the flowers is considered stagnant and may contain floating weed or leaves.

To add more to the fresh appearance of floral compositions it is customary to sprinkle the outside of flower vessels with spray, conveying the idea of dew. Metal vases should not be wetted, but nearly all pottery or porcelain vases are so treated after the flower arrangement has been made. The celadon vase is said to form an exception to this rule, because it becomes naturally covered with moisture condensed from the air.

FIG. 18

It often happens that a vase filled with water, but without any floral composition, is used in the alcove of a chamber. Such an arrangement is resorted to when there is a picture on the wall of the recess representing some flower of the season, in which case a few petals of the flower represented in the painting may be put into the water of the vase. A vessel simply containing water and no flowers, is an appropriate ornament for the chamber at a moon-viewing party, or when a picture of the full moon is displayed, the intention being to suggest the idea of the moon reflected in a lake. Sometimes a few maple leaves are placed in the vase in order further to sustain the idea of a natural sheet of water.

extolling their beauties as they are apt to restrict the imagination and detract from the fanciful sentiment conveyed in the poem.

Reference has been made elsewhere to the double associations from animal and vegetable life, used as favourite art motives by the Japanese. Such combinations are;—bamboos and sparrows, lions and peonies, nightingales and plum blossoms, deer among maples, wild-horses amid flowering grasses, and many others that might be mentioned. A combination of flowers and picture so as to produce a connected composition of this sort is considered very desirable. According to such a method of combination, a picture of deer requires in front of it an arrangement of maples, a painting of horses needs wild flowers, one of lions necessitates the use of peonies, and representations of dragons demand pine branches for the floral designs in the foreground. In the same way with figure paintings, when the figures represented are traditionally associated with particular trees or flowers, such flora should if possible be employed for the flower arrangements used before them. When for example, a picture, of Hotei, one of the Gods of Fortune, is displayed, bamboo branches should be disposed in front, and before pictures of Chinese children,—a common subject with Japanese painters,—coloured flowers are appropriate.

The connection of idea between the wall painting and the floral composition is occasionally one based upon the reputation of the painter or upon some fiction with which his name is associated. A famous Chinese painter called *Tô-Emmei*, whose works are greatly valued in this country, is said to have professed a great passion for chrysanthemums, hence, when a painting by this artist is displayed, it is customary to use chrysanthemums in the flower arrangement. Plum blossoms are reported to have been the special fancy of another great painter—*Rin-Nasei*,—and these flowers are therefore placed before his pictures.

This kind of combination is quite irrespective of the subject of the paintings, except in cases where other important rules would be violated. If the particular flowers required are represented in the paintings, it would then be an error to use the same natural flowers in front. Such a selection would not only be redundant, but would tend to detract from the excellence of the painting. It must also be remembered that the pictorial hangings in a Japanese room are frequently changed and are not displayed continuously throughout the year as is the custom with oil paintings in European rooms. They are quite as much an expression of the season and occasion as the floral compositions are. The contingency, therefore, of pictures requiring flowers which are out

damp the dais, as they contain an inner vessel which holds the water and the flower stems. Even such a simple object as the vase-tray or tablet has its fixed measurements, and a few fancy forms in the shape of open fans, circles, or segments, are not uncommon. Sometimes this flat tray is replaced by a small ornamented stand or table of carved or lacquered wood, intended to raise and give more importance to small flower arrangements. As previously stated, excepting in the case of the low tub-like vessels intended to suggest water scenery and used for water plants, the surface of the water and the fastener holding the stems in the vase, should be a little above the eye level of the seated spectator, so as not to be visible without effort.

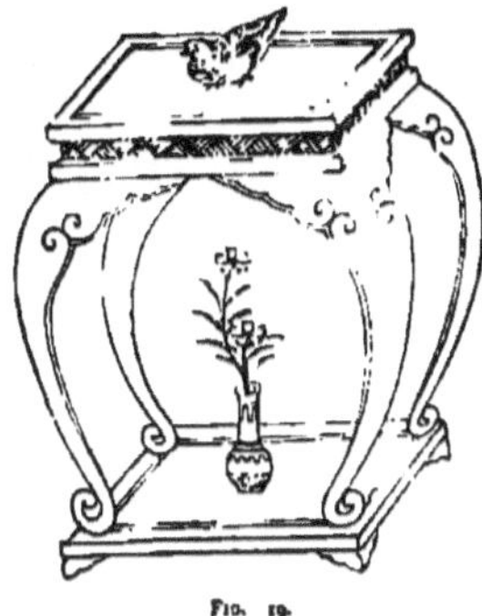

Fig. 19.

Some flower-stands exist which are of considerable height, having a bottom shelf. These are used for Incense Meetings, in which case the top of the table carries an ornamental incense burner, the shelf below being occupied by a very simple flower arrangement. (See Fig. 19). Examples of different trays and stands are shewn in Plate XXXI.b.

floral designs should occupy the dais of the *Toko-no-ma*, but for small and secondary arrangements the *Chigai-dana* or *Irregular Shelves* are often employed. As already explained in the case of the three openings of a high triple-mouthed vase, so for this tier of three shelves, triple arrangements of flowers should follow the natural distribution of growth observable in real scenery. For the top shelf, thick and moss-covered *tree* branches are chosen, for the middle shelf, young *tree* branches or land *plants* are selected, and for the lowest shelf, a composition with water *plants* should be employed. If *tree* branches be placed on the middle shelf, then the lowest shelf may have land *plants* instead of water *plants*. In the same recess, and in combination with the shelves, there is occasionally constructed an ornamental cupboard about eighteen inches or two feet in height. The paper slides of such cupboards are often painted with flower designs, and, in this case, care must be taken not to use natural flowers of a similar kind in vases placed on the adjacent shelves.

Strict rules are established as to the exact position in the *Toko-no-ma*, or principal recess, to be given to suspended vessels. First, with regard to *Boat-shaped* vessels, it is stated that in ancient times they were hung at a distance of about three feet from the ceiling or lintel, and exactly in the centre of the recess. In later times it became customary, however, to suspend them, at a distance equal to about a quarter of the span of the recess, from the corner pillar, which is always on the guests' side of the room, being that nearest the light. If, as is not uncommon, two *Boats* are suspended together, the upper one should be about sixteen inches below the lintel, and the lower one about the same distance above the floor of the recess; but this height is sometimes changed in order to suit the wall picture.

The direction given to the prows of such vessels, so as to suggest different ideas of motion, has been already explained. *Boats* are in some cases hung to the under sides of ornamental shelves or cupboards.

Well-buckets are not considered suitable for the recesses of important rooms, unless they happen to be a gift from a superior, in which case they may be given the place of honour; they are, however, frequently resorted to in second class rooms. When one bucket only is used, it is placed standing upon the floor of the recess, raised about fifteen inches from the surface by a stand of some sort, and removed about eleven inches from the corner pillar. Square buckets must be placed angle-wise. In the case of two suspended buckets, supposing the height from floor to lintel of the recess to be divided into three parts, the top of the lower bucket should be one third, and the top of the upper bucket two-thirds, of this height, from the dais. The bottom of the lower bucket thus becomes about eighteen inches from the floor of the recess, and is supported upon a small table, or on the edge of a stand representing a well-frame. The position of the well-pulley will be about one third of the span of the recess from the corner pillar, and the rope of the lower bucket being inclined and tort, it will be brought well to the side of the recess. If, as is usually the case, flat-sided buckets are used, one will have its sides parallel to the wall, and the other will present its angle to the front. Occasionally the above mentioned proportion as to height, is violated, and the lower bucket is placed immediately upon the floor, with the interposition only of a board. A style exists called the *Mossy-spring-buckets* in which this board is sprinkled with stones and moss. Sometimes a slab of decayed wood, or an irregular row of bamboo tubes resembling a Japanese drain-board, are used under the lower bucket. The *Piled-buckets*, consisting of two standing buckets, one supported on the edge of the other, are placed upon a flat drain-board immediately on the floor of the recess, and near to the corner pillar. One of these should be parallel to the wall and the other placed diagonally.

Hitherto flower compositions have been considered with reference to their disposition in the permanent recesses of chambers, either in the *Toko-no-ma*, or on the fixed shelves of the *Chigai-dana*. Other fancy arrangements, having no connection with the chamber recesses, also exist, though they are rare. Among such may be mentioned the arrangement of floral designs upon *Cabinets*, and *Flower-horses*.

Fig. 21.

The *Flower-horse* is so named because it is made in imitation of the ornamental clothes-horse or frame which is used in Japanese sleeping apartments, either for hanging garments upon, or for carrying large strips of rich curtaining serving as screens. The construction of such *Flower-horses* consists of two vertical and two horizontal bars of lacquered wood, framed together with the top bar projecting at the ends, the bottom bar being steadied and supported upon short cross-pieces. Such frames measure about five feet square. In the example given in Plate XXXVII., vessels of different kinds containing a variety of floral arrangements are hung to the side posts and cross-piece. From the centre is suspended a crescent-shaped bronze vessel containing an arrangement of small chrysanthemums and a trailing vine: from the right hand pillar are hung a cylindrical bamboo vase, with side mouth containing a drooping arrangement of *Patrinia scabiosæfolia*, and, below, a globular basket with a composition of carnations combined with the branch of some tree; and to the left hand pillar are fastened a hooked bronze vase with *Papaver rhœas*, and, below, a horn-shaped bamboo vase containing a double arrangement of *Platycodon grandiflorum* with barley.

There exist also fancy kinds of flower-stands, made of the *Flower-horse* combined with shelves and cabinets, and on which standing, hooked, and suspended flower compositions are arranged.

Flower-cabinets, in their simplest form, consist of two small shelves of different heights connected by vertical and horizontal framing, lacquered black, and ornamented with metal. For the upper shelf of such cabinets a drooping floral design is generally adopted, and for the lower shelf a standing arrangement. In the example shewn in Plate XXXVII., the top shelf supports a bronze vase with trailing ivy, and the lower shelf a porcelain vase containing asters. The same illustration shows a fancy cabinet called the *Thatched-kiosk*. It consist of a bottom board, with raised shelf supported on bamboo posts, and covered by a rustic thatch-roof. The shelf carries a bamboo vase from which springs a creeper, arranged to pass over the roof and hang down the side as a *Streamer*.

Other articles of furniture are sometimes pressed into service for the display of flowers. An example exists of a triple gong-frame, with one upper and two lower openings, from the cross bars of which cylindrical bamboo vases containing different kinds of peach blossoms are suspended.

A great fancy prevails for collecting and displaying together different kinds of blossoms of certain favourite trees and plants. The use of distinct varieties of the same blossom in one arrangement is opposed to the somewhat austere rules of the art of floral composition, but such specimens are sometimes displayed in separate vessels arranged on stands. The regent Hideyoshi is said to have devised an arrangement of five baskets of green bamboo disposed on the two shelves of a *Flower Stand*, and containing different varieties of the cherry blossom. Such arrangements, however, belong more to gardening than to chamber decoration.

PLATE XXXVII.

RUSTIC FLOWER CABINET

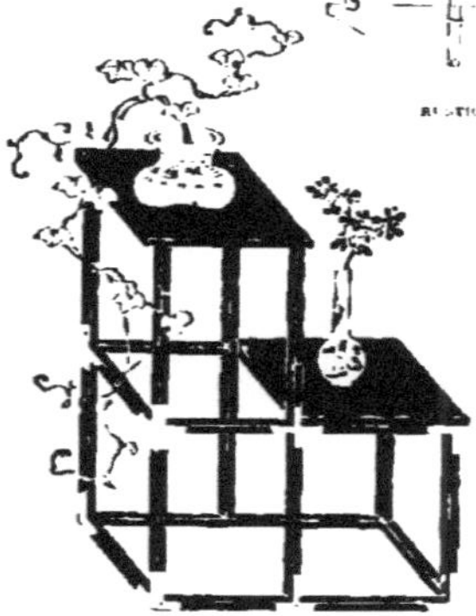

[illegible]

FLOWER [illegible]

Some writers go so far as to say that the floral design in a chamber should have a contrast in style with that of the adjoining garden. This fancy is better appreciated if it be remembered that during a great part of the year the outer walls of the Japanese house, which consist almost entirely of paper slides, are thrown completely open. If there be a landscape garden adjoining, consisting of lakes and hills, the floral arrangement in the room should by preference partake of a moorland character; but if the garden be level and waterless, then water plants or mountain trees should be selected for the flower decorations of the chamber interior.

CEREMONIAL AND ETIQUETTE.

THE Art of arranging flowers in Japan is essentially a polite accomplishment, and is governed by important rules and restrictions as regards etiquette and ceremonial. It is presumed that all floral designs are made mainly with the object of giving pleasure to visitors; and on certain ceremonial occasions they are actually intended to convey a silent compliment to the principal guest. Receptions given in rooms where flowers are arranged, often partake of the character of *Flower Meetings*, the guests in turn inspecting and admiring the host's floral design, or being called upon by him to make compositions of their own.

FIG. 22.

In attending such a reception, the visitor should leave his fan in the ante-chamber and, approaching within about three feet of the recess, seat himself in the old ceremonial attitude, with his knees bent and the body resting back on the heels. One hand should be placed on the knees, while the other respectfully touches the mats in front, the body being slightly bent forward. It must be remembered that there is always a supposed connection between the pictures which adorn the wall surface of the recess, and the floral arrangement standing or hanging in front. The guest should, therefore, first regard the *Kakemono* or picture, and if, as is often the case, there are three of these, he should examine first the central, then the left hand, and lastly the right hand one. Having thus bestowed his admiration upon the background of the scene, he may slide a little closer and inspect the floral composition in the foreground. In doing so he should first observe the central line of the flower arrangement, and then gradually examine left and right, and from top to bottom of the composition; lastly, with a word of apology for so doing, he may inspect the manner in which the stems are held at the bottom, this being one of the most difficult parts of the *technique*. It is considered impolite to put the face behind the branches and peer too closely into the flowers. After such inspection the guest slides a little back and regards the whole composition from a respectful distance, using some suitable expressions of admiration. Rules of etiquette actually go so far as to give the exact expressions to be employed in admiring different designs; for it is considered bad raste to apply indiscriminately exaggerated terms of praise not in keeping with the character of the particular flowers. When inspecting hanging floral arrangements the seated attitude should be changed for a standing and stooping posture.

A visitor is often invited to make an extemporary arrangement of flowers, for which purpose he is presented with certain suitable flower stems, or blossom-clad branches, and all the necessary utensils and implements. On such occasions the host must provide a vase, three quarters filled with water, which should be placed in the centre of the recess, upon a tray or table spread with a sheet of paper. In addition to this, a flower tray with two or three kinds of cut flowers, just as gathered with withered leaves and dead twigs left intact, a pair of scissors, a knife, a small saw, and a folded flower-cloth or duster, must be placed on the dais to the left of the vase, or in some convenient position on the floor of the chamber. The length, width, and manner of folding the flower-cloth, are all prescribed. Near to the above utensils and implements must be placed a jug full of water, and several forked twigs suitable for *Flower-fasteners*. These various tools and utensils are illustrated in Fig. 23. Others, such as a hammer, and plane, are sometimes employed.

Fig. 23

After asking a guest to arrange flowers, the host should commence to roll up

PLATE XXXVIII.

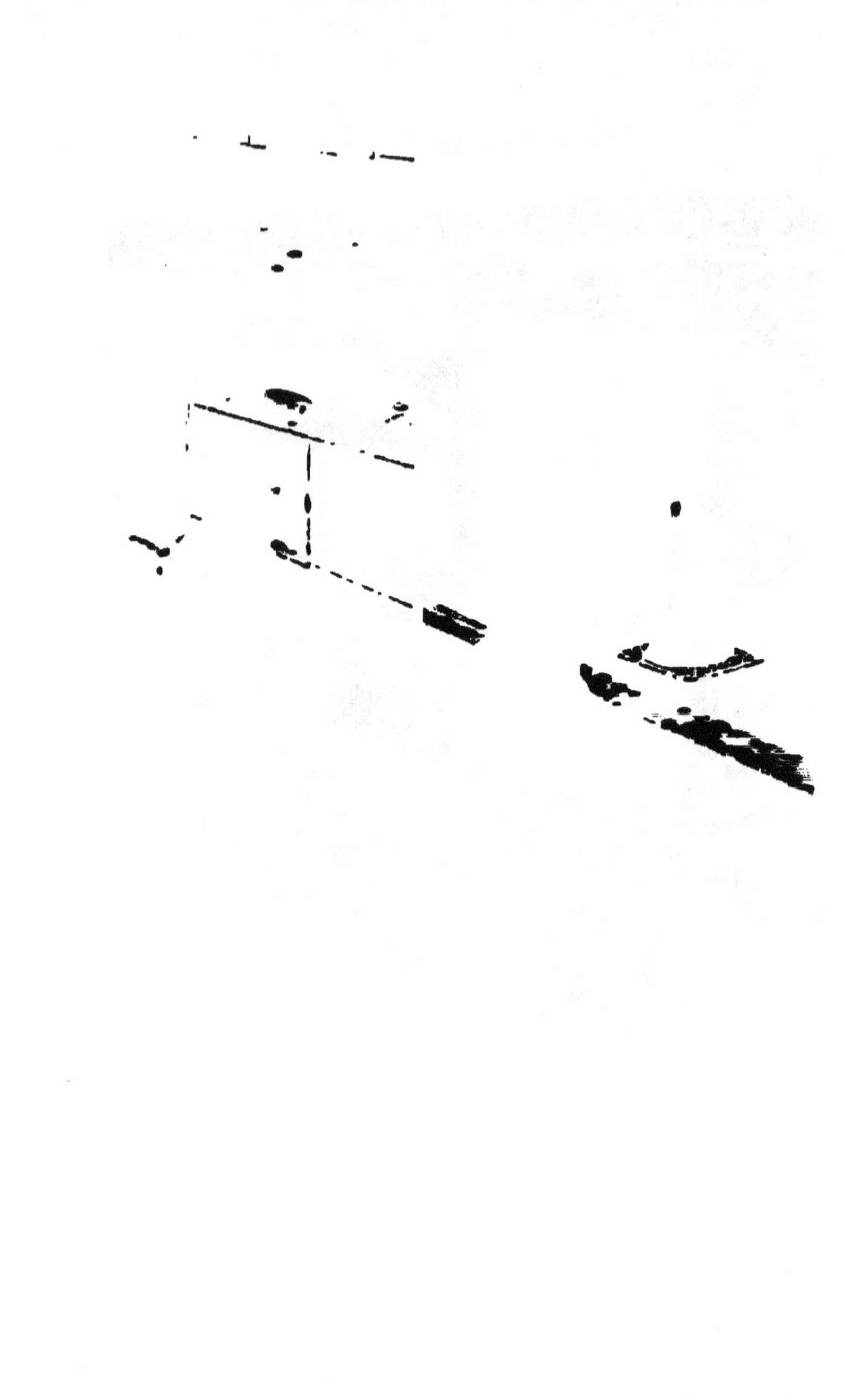

the *Kakemono* which adorns the recess, as it is considered to be demanding too much of a visitor to expect him to extemporize a flower arrangement in harmony with the picture which happens to decorate the recess at the time. The guest may, however, prevent the painting from being removed, thereby tacitly undertaking to make his flower composition accord with it. Should the master of the house produce a very rare and valuable vessel for holding the floral arrangement, it is polite for the guest to make objections, pleading want of sufficient skill to do justice to so precious a receptacle. If pressed, however, he must attempt a simple and unassuming arrangement of flowers, so as not to detract from the merit of the vessel itself. Should the host produce an insufficient quantity of cut flowers, the guest must make the best of them and on no account ask for more.

The visitor who is about to make a floral composition approaches the recess in which the flower vase is placed, and seats himself sideways towards it, facing the light, which in chambers of a superior kind is on the left side. He then fixes the flowers as quickly as possible, changing his position in order to regard them from a point immediately facing the recess, and altering and correcting them from this position. If the arrangement is intended to have a connection of idea with the hanging picture, it is placed fronting the edge of this picture, on the side nearest the light; but if no connection is intended, then it may be disposed centrally. Under no circumstances must the floral design project beyond the corner pillar of the recess. Having completed the composition, the designer should, as a matter of compliment, ask his entertainer to fill up the vase with water, and if this request be declined, he may then replenish it himself; he should not, however, press the host on this point, because correct judgment as to the amount of water suited for different arrangements requires considerable knowledge of the flower art, and it is possible that the host may decline on account of ignorance. When the water is filled in, the stand of the flower vase is wiped, and the different implements, with the exception of the scissors, are all put on the tray, and placed near the serving entrance, or right hand side of the chamber. The scissors are purposely left near the flowers as a silent and modest invitation to the master of the house to correct faults. The host, brings the dust-pan and brush, sweeps up any fallen leaves or litter, and removes the tray of tools.

If the hanging picture has been removed during the arranging of the flowers, the guest must now re-hang it, and see that the floral composition is placed so as not to clash with it in any way. When the whole arrangement is completed, the host and any other visitors present, who have meanwhile remained in an adjoining room, enter, and, approaching in turn the *Toko-no-ma* or recess in which the flowers are placed, salute and inspect in the manner previously described. The master of the house naturally confines his admiration to the floral arrangement, but the guests should also find compliments for the receptacle in which it is placed. After the Flower Meeting is concluded and the visitors retire, the guest who has arranged the flowers should remove them from the vase, placing them upon the board on which the vase stands, or on the wash-basin of the adjoining verandah; unless he is specially requested to let his work remain, it is considered presumptuous for him to quit without destroying the evidence of his skill (or clumsiness). This rule does not apply to visitors of very superior rank, who may be considered to show honour to their host by leaving their designs intact, however indifferent they may be.

Politeness during such meetings is considered so important that, if a rule happens to be violated through ignorance or incapacity on the part of the entertainer, the guest must try his best under the circumstances, and must do nothing to call attention to the error. As an example, may be given the instance of a host producing scented flowers for a visitor to arrange at an *Incense Meeting*, which is an occasion on which flowers with perfume are prohibited. The guest must in such a case use the flowers, removing the full-blown blossoms, and making an arrangement of buds alone in as simple and unassuming a manner as possible.

Fig. 21

In giving presents of cut flowers for the purpose of flower arrangements, they must not be trimmed, or they will look as if they had been previously used. The sender must however, in selecting them, consider how they are capable of combination into a floral composition, and must include stems and branches suitable for the accessory parts of the design. Such flowers should have the bottom of their stems placed in fancy paper wrappers. The form of paper wrapper suited to *tree* branches and that used for *plants* differs slightly. When flowers are offered as presents, the number of buds should exceed that of the open flowers, and they should be accompanied with withered or worm-

JAPANESE INTERIOR. ARRANGEMENT OF SPRING FLOWERS.

eaten leaves, and cobwebs, all left intact so as to look as if the [illegible] gathered. The recipient of such presents should carefully consider how these [illegible] best be arranged without injury or extensive alteration. If they appear to be quite unsuitable for a proper floral composition, it is better to place them in a vase as they are than to attempt a formal arrangement. In Fig. 24. is shown a branch of plum blossom held in a paper wrapper.

Presents of flowers are sometimes made to superiors, in a kind of hand-basket, which should be of white wood and quite new. Several sorts of flowers are generally placed in such vessels, which are furnished with lids perforated with four or eight square holes for the purpose. When such a present is received, it may be put on one of the shelves of a recess in a reception room, an empty vase being placed on the dais.

In a previous part of this work flowers considered specially felicitous during particular months have been enumerated. It remains, however, to consider, under the head of *Ceremonial*, certain floral arrangements which are fixed for important festivals. Of the numerous fête-days celebrated in Japan, may be mentioned first that of the New Year, and then the five great festivals called *Go-sekku*. The *Go-sekku* occur in all the months of the year of odd number, with the exception of the eleventh month, and, in four cases out of the five, fall on the day corresponding to the number of the month.

FLOWERS FOR THE NEW YEAR.

For the floral decorations of the New Year it is customary to adopt a combined arrangement of pine, bamboo, and plum branches, in a large bronze or porcelain vase. These *flowers* are, however, sometimes use separately, in which case the pine is displayed on the first, the bamboo on the second, and the plum on the third day of the year. In some cases a vase of green bamboo, with twigs and leaves left on, is used to hold branches of pine and plum trees, the floral triad being formed by including the vase itself. The willow is a favourite tree for use in hanging compositions at this season; and plants such as the *Adonis amurensis* (Fukujuso), *Rhodea japonica* (Omoto), and *Ardisia japonica* (Yabukoji), are often employed in combination with the plum.

FLOWERS FOR THE "FIVE FESTIVALS."

The [illegible] day of the New Year is the first of the Five Festiv[illegible] (Go[illegible] [illegible] [illegible] the Fete of the seven-plants (Nana-kusa). This festival is of Chi[illegible] origin, and its meaning is involved in some myster[illegible] it is undoubtedly connected with ideas of luck in the [illegible] ber seven. At its celebration seven different herbs [illegible] beaten with seven different household tools, and di[illegible] protection is invoked against the evils of the year. The flower arrangements adopted on this occasion do not diff[illegible] from those used for the first three days of the New Yea[illegible]

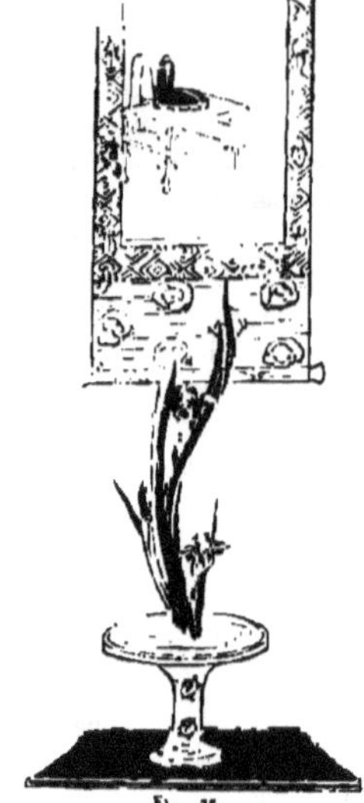

Fig. 25.

The third day of the third month is the second [illegible] the Five Festivals, and is called the Fête of the dolls. It is the national fête for girls, which they celebrate by displaying richly attired images representing the Emperor and Empress surrounded by courtiers and musicians. On this occasion peach blossoms, willow branches, cherry blossoms, the *Rhod[illegible] japonica*, and *Rosa indica*, are chosen for flower designs.

The third of the Five Festivals, falling on the fifth day of the fifth month, in the Fete for boys, called, by the Japanese, *Tango*. On this day, large painted paper fish made hollow so as to fill and float with the wind are placed on high bamboo poles in front of the dwellings, together with standards and other ornamental signs, each male child of a household being entitled to display at least one fish. The flower most honoured on this occasion is the flag or iris. A kind of early chrysanthemum, and a particular sort of bamboo, called the *Moso-chiku*, are also used.

The next festival falls on the seventh day of the seventh month and is called the Fête of *Tanabata*. It is the celebration of the meeting of two planets, and is supposed to be a day of very good omen. Slips of paper inscribed with poems and leaves of bamboo are hung up as ornaments on the occasion, and a general feasting takes place. The flowers used for compositions on this holiday should be *Platycodon grandiflorum* (Kikio), bamboo, and *Anthriscus arguens* (Karukaya), arranged together, or

PLATE XL.

Patrinia scabiosaefolia (Ominaeshi) employed [illegible]. A special [illegible] consisting of a couple of suspended bronze boats, called the *Facing Boats*, [illegible] then prow-face, and containing paper chrysanthemums, is used. The paper flowers should be of seven colours, and among them should be one real flower. To the *Streamer* of the floral design in one of the boats should be hung two, and to that of the other, five loops of silken cord of five different colours,—white, yellow, blue, purple, and red. These cords are called the *Worshipping-cords*, being associated with the idea of worship to the planets. A somewhat similar method of composition is adopted for standing vases, seven different flowers being placed in three vessels. The central vase contains three flowers, of which the *Principal* should be formed of a branch of the sacred tree *Cleyera japonica* (Sakaki), and the other two vases should each hold two flowers. The floral designs in all three vases must have their stems bound with the *Worshipping-cord.*

The last of the Five Festivals occurs on the ninth day of the ninth month, and is called the fête of chrysanthemums. Its origin is Chinese, and it is connected with a fancy that this flower imparts long life to those who imbibe an infusion of its blossoms in wine. For this occasion, as the name of the festival implies, the chrysanthemum is invariably used for floral arrangements. Flowers of five colours are employed for first-class compositions, white for the *Principal*, light red for the *Secondary*, dark red for the *Tertiary*, blue or purple for the *Sub-principal*, and yellow for the *Support*. Sometimes a design made with flowers of one colour is fancifully called the *three-colour* composition, the green of the leaves and the black of the bronze vase being included in the colours. The *Rhodea japonica* (Omoto), and *Nandina domestica* (Nanten), are also sometimes displayed on this occasion.

In addition to the Five Festivals just enumerated, there are other felicitous days in the year on which special rules are adopted for the flower arrangements.

For the first day of the sixth month, which is the beginning of the *Dog-days* or hottest season, water plants should be arranged in a sand basin or broad-mouthed vessel, together with white sand and pebbles, which are supposed to suggest ice.

For the first day of the eighth month, called the *Hassaku*, the flowers used should have fruit or berries, and all faded leaves or branches must be carefully removed. This fête was one specially honoured by Iyeyasu, the first of the Tokugawa Shoguns, to celebrate his entrance into Yedo.

On the eighth day of the eighth month white flowers should be displayed.

For the harvest festival of the fifteenth day of the eighth month, flowers of the season should be employed, with the addition of a sheaf of the ripe rice-plant, in celebration of the harvest.

The above special rules, though hardly applicable to any European adaptation of the art under consideration, are interesting as showing how a certain harmony of idea should always be sought between the floral design and the special occasion celebrated. The following are ceremonies which may occur at any period of the year, with general rules for their appropriate flower arrangements.

FLOWERS USED AT BETROTHALS.

On this occasion visits take place and presents are interchanged. The flower or flowers specially felicitous for the particular season, in accordance with the list previously given, must be used. The iris is considered specially suited for this purpose, unless out of season. Some people, however, object to the use of flowers of purple colour.

FLOWERS FOR WEDDING FESTIVITIES.

With regard to flowers arranged for weddings, it must be remembered that, amongst colours, red is regarded as *male* and white as *female*. Hence in the case of a son-in-law being adopted into the family of his bride, the bridegroom being regarded as the guest of the occasion, the *Principal* line of the floral design should be red, whilst the *Supporting* line is of the *female* colour—white. On the other hand, when a bride is adopted into the family of her husband, she being considered the guest of the occasion, the white colour has the central position in the arrangement. In both cases the stems of the flowers used must be closed, and firmly connected at the base with coloured cords, called *Mizuhiki*, to signify union. The bamboo and pine, being always in season and specially felicitous, are considered well suited for wedding ceremonies. They should be placed separately in a pair of similar vases, the pine towards the guest's side of the chamber, and the bamboo towards the host's side. Purple flowers, willow branches, and other drooping plants are prohibited for

JAPANESE INTERIOR LADY ARRANGING LOTUS FLOWERS.

weddings, as also are all floral compositions placed in suspended vessels. Purple is considered expressive of mourning, and all loose drooping arrangements imply disunion and inconstancy

FLOWERS FOR COMING OF AGE CELEBRATIONS.

There were formerly two coming of age celebrations for males one on the occasion of the boy first assuming the *hakama* or ceremonial trousers, and the second when his long locks of hair were cut off and the youth assumed the cue worn by adults. For both of the above festivals, flower arrangements were required to be firm and vigorous, with a large proportion of buds and young branches. Faded branches and full blown flowers were prohibited.

FLOWERS FOR PROMOTIONS IN RANK.

Flowers used on the occasion of honourable promotions should have buds below and open flowers above, to signify ascent in rank, and the use of withered leaves or over-blown flowers must be carefully avoided.

FLOWERS FOR THE CEREMONY OF RELIGIOUS RETIREMENT.

Upon reaching a ripe age it has always been the custom in Japan to seek retirement from active life, the head being shaven and a semi-religious ceremony performed. Floral designs used at such a celebration should consist of berry-bearing plants or trees, and red flowers must not be employed.

FLOWERS FOR OLD AGE CELEBRATIONS.

Old moss-clad branches, flowers late in season, and second flowerings, are suited to such occasions. Withered branches and flowers which quickly fade or fall to pieces must on no account be used.

FLOWERS AT FAREWELL GATHERINGS.

There are certain flowers which bloom twice in the same year, and these are technically called *Returning-flowers*. Though considered unsuitable for ordinary occasions, because out of harmony with the season, such flowers are specially appropriate for farewell gatherings. The idea in so employing them is to express the hope for a safe return. Similar arrangements were adopted at feasts before a battle or campaign.

FLOWERS FOR THE SICK.

Flowers placed before sick persons should be put together in a rapid and unlaboured manner, and should be vigorous in style, to suggest the idea of recovery and strength. At the ceremony of praying for the sick, flower compositions should be full and gay, as well as bold and powerful in style. The use of the pine, *Podocarpus macrophylla*, and *Rhodea japonica* should be avoided.

FLOWERS BEFORE HOUSEHOLD SHRINES.

Each household in Japan has generally two shrines, one dedicated to the household gods of the old Shinto cult—the national deities of the country,—and the other to the spirits of deceased relatives, which is Buddhist. For arrangements of flowers before the Shinto shrine, a full and powerful composition is required. All ugly flowers, those of strong odour, and those having thorns are prohibited. A special branch called the *Facing-branch* is used behind the *Principal* line of the composition and pointing towards the shrine. The floral design placed in front of a Buddhist shrine should also be full and crowded, but the arrangement should not be laboured. The front of the design must face the shrine and not the spectator. A story is told of a Daimio who went to his ancestral mortuary chapel and arranged flowers with great skill and labour much to the admiration of the chief priest, who, however, pointedly asked him whether he had made his composition with a desire for public praise or in reverence for the spirits of the dead. Unable to answer, and struck with the justice of the reproof, the noble altered the flower arrangement to one more simple and unassuming. White flowers are specially suited for such designs, and in all cases, as little artificial trimming and bending as possible should be resorted to, the withered leaves only being removed.

FLOWERS AT DEATH ANNIVERSARIES.

For floral designs from the first to the fiftieth anniversary of a death, a quiet and simple style must be adopted. White and yellow flowers are used, and with them is combined a branch of the sacred tree *Cleyera japonica* in token of a religious offering. Sometimes a withered branch is employed for the *Principal.* As such flower arrangements must be unassuming, and the composer must not attempt to display skill, it is considered ignorant and in bad taste to criticise them, judging them by the ordinary standards. The selection of flower buds is to be avoided, full-blown or over-blown flowers being preferred. The use of crossing leaves, or flowers blooming twice in the year, is also forbidden.

On and after the fiftieth anniversary, gayer arrangements and even red flowers may be employed, and the *Principal* line should be formed with an old moss-covered branch, a flower of the season being added as an auxiliary. Flowers of ominous names, such as the *Oni-yuri* (demon-lily), must on no account be used.

FLOWERS AT PRAYERS FOR RAIN OR FINE WEATHER.

There are special rules with regard to flowers which are suited for the occasions, of prayer for rain or fine weather. As it is the east wind which brings rain, floral arrangements used at time of praying for rain should have their *Principal* line pointing from right to left, to suggest the east wind blowing. A reverse arrangement is resorted to on occasions of prayer for fine weather, when the *Principal* line, leaning from left to right, is made to suggest the west wind.

FLOWERS USED AT MOON VIEWING.

Moon viewing is at all times a favourite pastime of the Japanese, but the great moon festival of the year is on the fifteenth day of the eighth month. The more important dwellings have a special chamber with open galleries from which the sight of the moon-lit landscape can be enjoyed. The floral arrangement occupies the recess of the chamber and has of course no real connection with the outside prospect, but in the flower composition itself the idea of a moon-lit landscape is expressed. A branch of a pine tree is used, and between the *Principal* and *Secondary* lines of the composition a special branch is introduced, fancifully called the *Moon-shadow-branch*; a hollow gap also is formed between the foliage, bounded by a special branch called the *Dividing-branch.* In the composition, the idea is to suggest both the opening through which the moon can be partially observed, and the dark branch which appears to cross its surface. To fully appreciate the analogy one must be familiar with the scenery of Japan, and have seen, on a clear night, the irregular pine trees standing out against the moon-lit heavens.

FLOWERS FOR HOUSE WARMINGS.

The Japanese word for red (*Hi*) is phonetically the same as that which denotes fire, and as conflagration is the dreaded enemy of the beautiful but perishable buildings of the country, a superstitious objection exists to the use of red flowers on the occasion of such celebrations as house-warmings. Flowers the names of which contain the ominous ideograph *Hi* are also prohibited. Such flowers are, the *Helianthus* (Himawari), *Lychnis grandiflora* (Gampi=Gan-hi), *Lilium concolor* (Hime-yuri), and the *Thuya obtusa* (Hinoki). The use of the last, (Hinoki), is specially disliked, as the wood of this tree was anciently employed for producing fire. By a curious perversion, however, some Schools approve and even favour the use of *Hinoki*, because the name is phonetically synonymous with another word *hinoki* meaning protection from fire.

For floral designs on occasions of this kind, a two-story bamboo standing-vase is suitable, having water plants arranged in both openings, the reason given for such an arrangement being that water is the extinguisher of fire. Another favourite combination in Summer is that of a white magnolia branch with white irises. In Winter, suitable flowers are scarce, and the narcissus is used in the lower opening of the vase, with water alone in the top.

FLOWERS AT POETRY MEETINGS.

A favourite pastime of the educated classes in Japan in former times, was the meeting of friends for the composition of verses. Sometimes a flower of the season formed the subject of competitive versification, and in this case the Enshiu School held that the flower chosen as a theme, if in season, should be displayed in a simple and quiet arrangement in the chamber of meeting. Some Schools were, however, of a contrary way of thinking, maintaining that the particular flowers selected as subjects for poetical composition should not be displayed at the meeting.

FLOWERS FOR THE TEA CEREMONIAL.

As the chambers in which the tea ceremonial is conducted are of very limited size, the flower arrangements must be small in scale, simple, and unaffected.

The displaying of the floral design has its proper order in the tea ceremony. It is not generally placed in the recess at the same time as the hanging picture. First, some suitable *kakemono* or painting is shown, and a part of the entertainment takes place, after which the guests retire; the *kakemono* is then removed, and the flower arrangement is prepared before their re-admission. Partly to avoid keeping the visitors waiting too long, and partly on account of the severe simplicity of the surroundings, and the austere character of the entertainment, the flower composition is made as unassuming as possible. The kind of floral design adopted for tea rooms, differs, therefore, from the ordinary style, being less elaborate and studied: it is distinguished by the name of the *Thrown-in-style*. If a standing vase be employed, the composition consists of the *Principal* line used singly; in a hooked vase, the *Secondary* line alone is used; and for a suspended vessel, the *Tertiary* line is preferred. If three stems are combined in one design, they must not be divided, but must be kept together in one line. Flowers used on such occasions must be well sprinkled with spray, to look fresh and as if covered with dew. When a tea ceremony takes place in a large room, instead of in the orthodox miniature chamber, then both the flowers and mural picture may be displayed at one time. If such meetings are held at night, floral arrangements are not employed. A hooked or suspended arrangement of flowers is preferred for tea rooms. Red blossoms, and flowers having strong perfumes, are disliked. The following flowers are also objected to:—*Patrinia scabiosaefolia*, cherry, rose, lotus, *Cnicus*, *Calendula officinalis*, *Nuphar japonicum*, *Celosia argentea*, *Illicium religiosum*, orchid, pomegranate, daphne, *Lespedeza*, drooping lily, drooping *Platycodon*, tea plant, and *Anthistiria arguens*.

One of the objects sought in arranging flowers in a tea room is to arouse the admiration of the guests by the quaintness and novelty, as well as the unassuming simplicity, of the composition. For this reason, curious and unusual vessels are often resorted to.

FLOWERS FOR INCENSE MEETINGS.

Another polite pastime of the Japanese in former times was that of burning different kinds of incense, and of distinguishing the scents produced; a small room like the tea room being generally used on such occasions. The little bronze or porcelain incense-burner became the principal art object of the chamber, and was supported on a stand or low table, about one foot high, placed on the dais of the recess. The height of this stand necessitated that the hanging picture should be a broad low one, technically called a *Cross-piece* (Yoko-mono). The most simple kind of floral decoration at these entertainments was that of a single flower and leaf, or a small bunch of flowers, placed in a tiny vase below the stand or table. (See Fig. 19). Some theorists object to this arrangement because the incense-burner above contains fire, and both wind and fire are injurious to flowers. If not used in this manner, the flowers should be placed in a vase hooked to the pillar or wall of the recess, or a suspended boat arrangement may be adopted. All scented flowers of whatever kind are prohibited at incense-burning meeting. This rule excludes such flowers as the plum, daphne, chrysanthemum, and *Aconitum fischeri.*

In cases where a hanging incense-burner is suspended from the lintel of the alcove, a standing arrangement of flowers may be placed below.

MANIPULATION.

PREPARATORY to the wedging of plant and tree cuttings into vases and other receptacles, certain precautions are necessary in order to prolong the vitality of the vegetation thus employed. For, however well designed a flower arrangement may be, unless it possesses a certain degree of vitality and endurance, it will speedily lose its shape, and become limp, drooping, and faded. It is important that cuttings from trees and plants should be made at dawn or dusk, except in the very coldest season. If not manipulated at once, they should be kept in a cool damp place until required, laid if possible upon stone or cement paving. They may with advantage be suspended down the hollow of a deep well, but not so as to dip into the water. Large delicate blossoms or leaves are wrapped in soft paper, and stems are loosely bound together with soft cord to prevent injury. After keeping in this manner for some hours, and sometimes for a whole day, the cuttings, just before use, are laid floating in tubs or bowls of clean cold water. During the heat of Summer, and at such a period only, this floating in water takes place immediately after cutting. In temperate and cool seasons, however, immediate immersion in cold water is found to be injurious to long vitality, and productive of brittleness in the stems, which have to undergo considerable bending during their preparation for combining into a flower arrangement. In the cold and frosty weather, floral cuttings are not laid in water, but simply sprinkled a little and kept in a shady place. The stems and branches are at such season very brittle and difficult to bend without breaking, and it is usual to slowly warm and thaw them over a charcoal brazier, in order to make them softer and more pliant.

The two essential points to be secured with cuttings for flower arrangements, are, first,—to preserve the succulency of the stem extremities, so that the water from the vase continues to rise into their fibres; and second,—to obtain softness and pliability in the stems and branches enabling them to yield to the twisting and bending which is necessary in order to give them the required curves.

The cut produced by scissors or knives in removing branches or flower stems from the parent growth is not found to be conducive to the long vitality of such cuttings, when wedged into vases containing water. Without professing to explain this scientifically, it may be surmised that the surface of such a section given to a branch or stem has a tendency to close and heal, thus choking the pores by which it is necessary that the water should rise. Some growths are of course more hardy than others, but with many, an artificial treatment of the extremities of the cuttings, before wedging them into their receptacles, is found to considerably prolong vitality.

The stems of the iris, chrysanthemum, and some other plants, have their ends charred by fire to increase their succulence. In the case of the bamboo, wistaria, rush, monochoria, lichnis, and other growths having knotted stems, this method is found to be useless. Charring is, moreover, never resorted to in the Summer time. The stems of some plants,—the peony for an example,—have their ends plunged in boiling water, containing sulphur; or sometimes a little arsenic is used instead of sulphur. A method followed by some masters, is that of mashing or crushing the ends of the cuttings, using the back of the flower-scissors for the purpose; this is the manner invariably adopted by the Ikenobo School. Whether the extremities are burnt, boiled, or crushed, the cuttings so treated are afterwards held upside down, and a shower of water poured over them for some minutes. Bamboo tubes are specially treated to keep them green and fresh. They are progged through the centre with a squewer so as to pierce all the intermediary knots or divisions, leaving the bottom one only intact, and then water is filled into the tube. Sometimes a dilution of boiled cloves or of rice wine is added to the water as a stimulant.

The artificial curves imparted to branches in Japanese compositions are produced by careful bending, given with the two thumbs and fore fingers placed close together, and with a force just short of breaking. Some stems of more yielding character will readily assume the required form without snapping, but the harder and more brittle kinds are variously treated. They are sometimes shaved with a knife at the points to be bent, or are softened with boiling water, or heated over a slow fire. A cloth or rag is wrapped round the part to be bent, to prevent splintering. If dipped in water after such forcible bending the elasticity is destroyed and the branch is prevented from returning to its original shape. To avoid ugly angles it follows that several bends have to be made in a branch to obtain the required curve.

Means are resorted to in order to keep large heavy blossoms in position in compositions and to prevent them drooping or falling off. Camellia flowers quickly fall, and it is found that keeping damp the junction of the blossoms with their stems by means of salt prevents this. With peonies, large chrysanthemums, magnolias, sunflowers, and other large blossoms, which play an important and fixed part in compositions, their exact number, position, and direction being fixed, it is often necessary to resort to the artificial support of hidden bamboo spikes or wires, though these are tricks not encouraged by the masters, for fear of being abused. Even painting is secretly resorted to in some cases to preserve the apparent freshness of colour in flowers and leaves, or to give them the appearance of other more beautiful specimens. The writer has heard of red camellia flowers in compositions being dotted with sulphur to give the appearance of the variegated kind of blossom. The green pine needles, which quickly turn brown and lose their gloss, are often washed with a mixture of gamboge or sulphur and size. The bark of tree stumps is also sometimes painted in a similar way.

Methods which are strictly speaking in violation of the principles of the art as laid down, are often resorted to. One of these is the use of *Borrowed leaves*, by which is meant leaves of one plant used to embellish the flowers of another plant possessing somewhat similar leaves which are not in condition at the time. There must be some resemblance, however, between the real leaves and the substituted material. The leaves of the young oak tree are in this way sometimes added to rhododendron flowers, the real leaves of which wither and fall before the blossoms.

In like manner *Borrowed flowers* are sometimes added to handsome leaves of a different plant the flowers of which are in poor condition. Thus, to the leaves of the camellia are occasionaly added flowers of the *Hibiscus syriacus*, and to chrysanthemum leaves, the flowers of the *Inula britannica* are sometimes united. The resemblance here is between the flowers and not between the leaves of the two growths that are combined.

Lichen and moss are produced upon branches and stumps of tree cuttings by artificial means. For this purpose the branches are placed on a tile roof exposed to the wet, or kept on the shady side of a house in a damp spot. Moss is considered desirable on the bark of the pine, fir, plum, cherry, cypress, and azalia, when thick branches or stubs are used; and lichen is considered suitable for the pine, fir, oak, maple, and willow.

PRACTICAL EXAMPLES.

THE following illustrations, from Plate XLII. to Plate XLIV. inclusive, are given to aid the designer by showing in juxtaposition defective and corrected arrangements of the same floral branches or leaves.

In Plate XLII. the *Kerria japonica* is represented at A, placed in a standing vase, with its branches very much in the form assumed when first gathered. The defects of this composition are, the stiffness of the central stem, the weak and open appearance of the branches at the base, the redundancy and parallelism of some of the smaller sprays, and the general fulness, irregularity, and formlessness of the whole. Altered as in B, the central stem or *Principal* assumes a stronger curve, the different branches are united at the base, the lines of the *Secondary* and *Tertiary* are improved, and some of the smaller sprays removed, so that the whole becomes disposed into five governing lines.

In Plate XLIII. the *Spiraea cantoniensis* is shown at A, carelessly arranged in a hooked bamboo vase, having a side mouth. Here the central line is too long for a hanging arrangement, which requires the drooping branch or *Streamer* to be the longest, and the branches are too numerous and straggling. Altered as at B, the *Principal* is shortened, the *Secondary*, being in this case a *Streamer*, is bent down in a wave-like curve, some of the excessive sprays are removed, and the whole resolves itself into a three-lined arrangement.

In Plate XLIV. the leaf-orchid is shown, at A, in a defective and almost symmetrical arrangement. The principal faults of this composition, are the straightness of the central leaf, and the monotonous manner in which the fronts and backs of the leaves, as shown by the shading, are disposed. Improved, as at B, the stem of the central leaf is gracefully curved, showing its front surface with the edge curled over, and the other leaves are united firmly at their base, arranged in steps, and disposed so as to reveal alternately front and back surfaces in a well balanced variety. The whole assumes the character of a tri-lineal design.

The special rules and traditions which govern the arrangement of particular flowers will now be explained, practical examples being shown, so far as the scope of this work permits, in the accompanying illustrations.

ARRANGEMENT OF PLUM BRANCHES.

The plum blossom being the earliest flower of the year, is held in high esteem for floral arrangements. The hardiness of the plum tree, the duration of its blossom, its sweet perfume, and the austere type of its beauty, all help to make it even a greater favourite for flower compositions than its more showy rival the cherry tree. There exist in Japan many varieties of this blossom,—red, white, single, and double. Branches of red plum blossoms should be arranged in a fuller and gayer manner than those of white blossom, which should be displayed in a quiet and open style. For Spring arrangements a fancy prevails of suggesting, by means of a perch-like bend of a branch in the composition, the presence of the *Uguisu*, sometimes called the Japanese nightingale, a bird which delights to make its home of song in the plum trees. Heavy and antique standing vases are the most suitable receptacles for this flower.

The rough and irregular character of the branches of this tree, renders considerable trimming and bending necessary in order to arrange them into lineal compositions, and this manipulation is assisted by softening refractory portions in hot water. In Plate XLV. is illustrated the process of composing a tri-lineal design with small branches, each stem being shown both in its natural and its altered form, side by side with the completed design.

But the favourite kind of plum branch for chamber decoration is that of the thick, rugged, moss-covered trunk, with young branches and shoots attached. The fresh shoots, which grow vertically in nearly parallel lines, are specially admired when combined with the rugged trunk, and are introduced under a special name, (*Zuwaye*), taking their appropriate place in compositions. Arrangements or this kind vary according to the class of room in which they are used. For important chambers a composition leaning to the right is adopted, and for secondary rooms one leaning to the left. Such designs are said to have respect-

PLATE XLII.

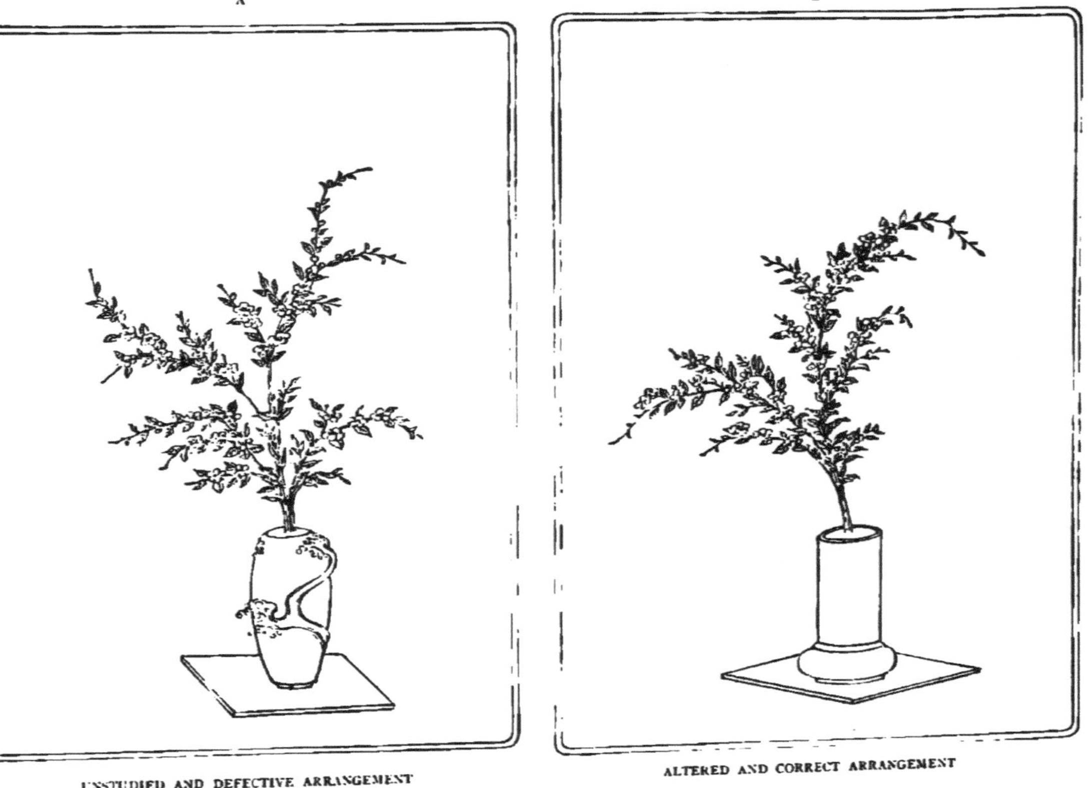

A — UNSTUDIED AND DEFECTIVE ARRANGEMENT

B — ALTERED AND CORRECT ARRANGEMENT

PLATE XLIII.

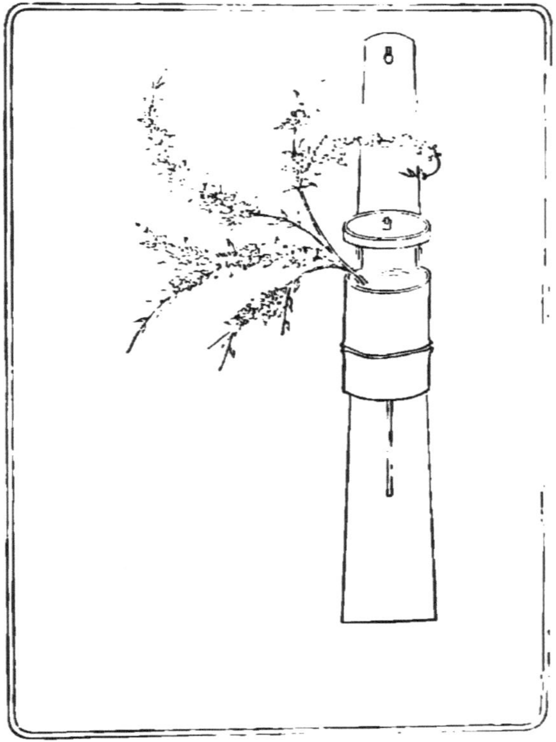

UNSTUDIED AND DEFECTIVE ARRANGEMENT.

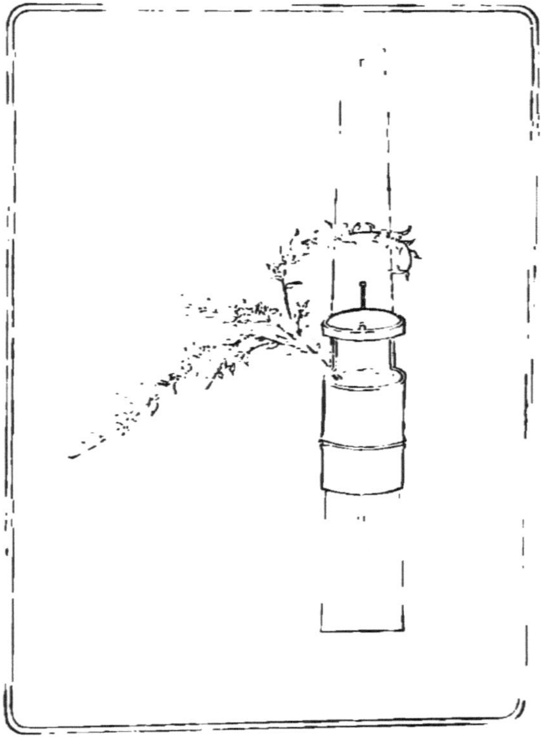

ALTERED AND CORRECT ARRANGEMENT.

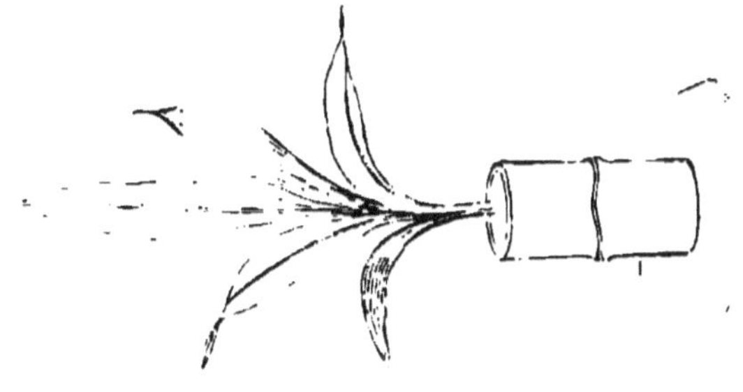

ively *South* and *North* character. For arrangements of *right* or *South* character, full and half-open blossoms predominate, buds being employed sparingly in the *Tertiary* branch only, whilst for those of *left* or *North* character, no full blossoms are used, but only buds and half-open flowers, a more quiet and bare character being given to the whole composition. The shoots are, in both designs, sparingly introduced, springing from the main trunk in three parallel lines, and in a position between the *Secondary* and *Tertiary*. In this particular, the plum tree forms a privileged exception to the general rule prohibiting *parallelism* in the disposition of stems or branches.

A curious arrangement of plum branches in a large shallow basin filled with water, illustrated in Plate XLVIa., requires special notice. As a general rule these broad water-basins are employed only during Summer, and for the display of trees or plants which grow near or in water. In the present case, the composition consists of an irregularly bent cutting of a plum tree disposed in an oblong shallow vessel, in the *hanging* style, with the principal branch diving through the water and re-appearing beyond. The explanation of this fanciful arrangement is that it was originally in imitation of the famous Recumbent-Dragon-Plum Tree at Kameido, remarkable for its crawling trunk which ploughed the ground in several places before re-appearing clad with blossoms. In the artificial composition it is considered important that the extremity issuing from the water should rise firmly, to suggest springing from the earth, and should, in fact, be fixed as if it were a branch separately held below, though sufficiently conveying the idea of continuity of line with the original branch. Some Schools otherwise explain the origin of this style of design. By them it is called the plum of the mountain stream, or the water-diving plum, and its introduction as a floral design is attributed to the æsthete Soho, who, during a hunting expedition, observed a crooked old plum tree dipping in this manner into a mountain stream, and adopted the idea for application in the flower art which he practised. A tree of this kind, measuring over a hundred feet from its root to the extremity of its branches, is said to have existed near Kioto. In the above arrangements, sand and gravel are mostly used for holding the plum branch in position, but sometimes the *horse's-bit fastener* is employed; in which case, for reasons already stated, the composition must not be placed on the dais of a chamber of importance. In one of the illustrations of chamber interiors, (Plate XXXVIII.), is shown a composition with plum branches in a bamboo vase containing also narcissus flowers below. The clothing of the figures indicates the early Japanese Spring,—the season of the plum blossom.

The plum is used in combination with land and water plants such as the Spring chrysanthemum and narcissus. (See Plate XLVII.A). An elaborate composition in a large bamboo vase with five mouths is sometimes seen, having a branch of white plum in the top, a willow branch below, then a narcissus, after which a camellia, and lowest of all a pine branch. In Plate XLVI. may be seen a branch of plum blossoms arranged in a hanging basket, in combination with the plant *Adonis amurensis*.

ARRANGEMENTS OF PINE BRANCHES.

The pine is the most important of all flowerless trees in Japan. The hardiest and noblest of evergreens, and a constant feature of the landscape,—whether it be mountain or coast scenery,—bent and twisted by the wind into shapes so quaint and contorted that faithful imitations in miniature may well be mistaken for grotesque caricatures, this venerable tree forms a favourite subject for poets and artists, and finds an important place in floral compositions on all congratulatory occasions. Combined with the plum and bamboo, and associated with the crane and tortoise, it is used in decoration to express the sentiment of happy old age.

The principal kinds of pine are the *Pinus thunbergii*, known by the Japanese as the *black* or *male* pine (Kuro-matsu or O-matsu), the *Pinus densiflora*, called the *red* or *female* pine (Aka-matsu or Me-matsu), and the *Pinus parviflora* (Goyo-no-matsu). The *Pinus parviflora*, on account of the straightness and delicacy of its leaves is often arranged in a simple vertical style, using thin sprays; but for compositions with other species of this tree, thick gnarled branches are preferred, and a bold rugged character given to the design. The pine is used mostly for arrangements in standing vessels, for which it is said to be specially suited, but hanging compositions are occasionally employed. A favourite treatment is that of a broad stump cut off horizontally, with a thick twisted branch springing from its base. At all times the pine branch is used as much as possible in its natural state, being trimmed until its arched foliage assumes a balance of masses suggestive of the three radical divisions employed in flower compositions, and the building together of separate branches, common with other trees, is comparatively little resorted to. An example illustrating the pine foliage disposed in cloud-shaped masses may be seen in Fig. 3, page 55.

The occasions for which the use of the pine is specially appropriate are at the New Year, at weddings, at old-age celebrations, and sometimes, though rarely,

PLATE XLV.

A

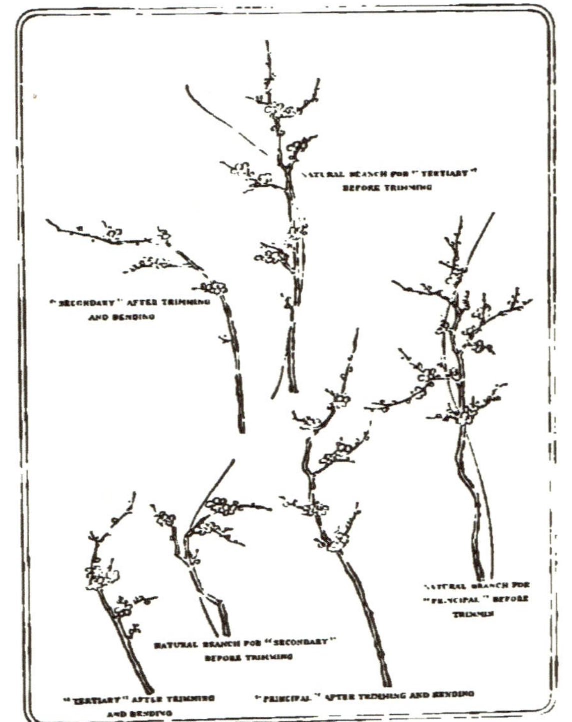

NATURAL AND ALTERED FORMS OF BRANCHES FOR "PRINCIPAL" "SECONDARY" AND "TERTIARY" LINES.

B

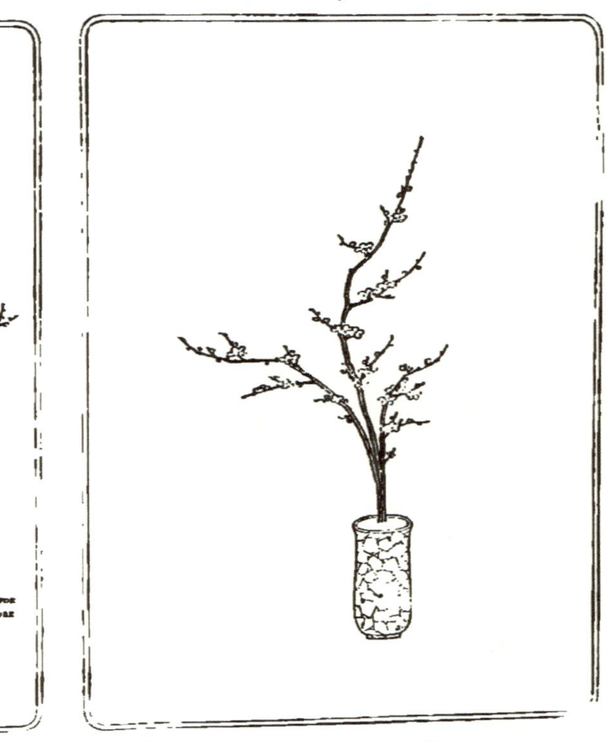

COMPLETED ARRANGEMENT.

PLATE XLVI

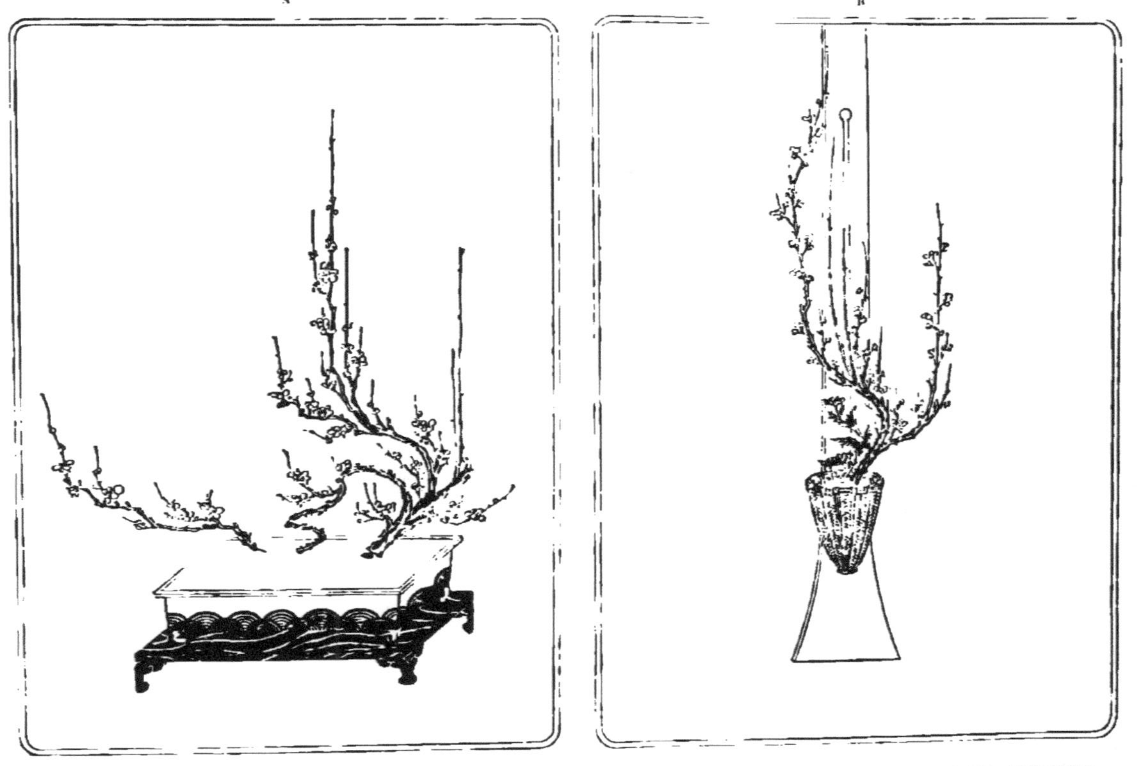

ARRANGEMENT OF WHITE PLUM IN BRONZE BASIN, WITH BRANCH DIVING UNDER THE WATER.

ARRANGEMENT OF PLUM BRANCH AND *ADONIS AMURENSIS* (FUKUJUSO) IN BASKET, HOOKED AGAINST PILLAR TABLET.

at farewell gatherings. At wedding feasts a double arrangement [illegible] of [illegible] standing vases is employed. For this purpose a branch of the [illegible] pine is [illegible] in one vessel, and a branch of the *female* pine in the other. The [illegible] of each design should be similar, but the branch of the *female* pine from the [illegible] vase should stretch a little beneath the corresponding branch of the [illegible] pine. These together are called the *Destiny-uniting* branches, and the complete design is said to typify eternal union. The same sentiment is expressed by arranging a branch of each of these trees, one below the other, in a single vase.

Occasionally, in suspended arrangements of pine, long stiff threads are hung from the branches, in conventional imitation of the parasitic grasses which attach themselves to this tree; and in disposing such threads, their balance into groups of three, five, or seven irregular lengths is carefully attended to.

The white chrysanthemum is sometimes used in combination with the pine, a custom introduced by the philosopher Rikiu, in allusion to a favourite verse of poetry comparing the white flowers of the wild chrysanthemum when seen beneath the pine branches to the moon between black clouds; the foliated leaf-masses of the Japanese pine tree piled one above the other, being not unlike in shape to rolling masses of dark clouds. This is one of the many examples in which the sentiment of landscape is introduced into flower designs. Another example is that of what is called the *Mountain valley and stream* composition, suited to a vase of three openings. In the upper mouth is placed a pine branch, to suggest a mountain summit, a land plant is placed in the centre, to suggest the plain or valley, and in the lowest mouth is arranged a water plant, to convey the idea of a stream or river. High bamboo vases are employed for such arrangements. Somewhat similar to the last named composition, is a design in a vase of two mouths with pine branches in each, the top branch being kept small in order to suggest a distant tree on the hill-top, and the lower branch full and elaborate to represent a tree in the foreground.

The use of the pine at moon-viewing gatherings has been alluded to in an earlier part of this work. Its employment in a single arrangement at farewell-gatherings was invented by Rikiu, being suggested to him by a verse of poetry in which the Japanese name for pine (*matsu*) conveys a double meaning, the same word also implying *waiting*, for a lover's return. Many of the traditions of the Flower Art are based upon such poetical allusions.

Examples, though rare, may be seen of the pine arranged in a water-basin in combination with a water-plant such as the iris or narcissus. One form of fancy arrangement, called the *Fuji-pine*, after the famous volcanic mountain Fuji-san, remains to be mentioned. In such a composition a branch is bent to resemble the outline of Mount Fuji, and is combined with the other branches and foliage in such a manner as to give the profile of the bare conical peak, and suggest at the same time the wooded country at its base.

ARRANGEMENT OF BAMBOOS.

In the Flower Art, the bamboo is, strictly speaking, regarded neither as a *tree* nor a *plant*, but it may occupy the position of either. It should never be placed in a vessel made of bamboo. In combination with other trees or plants, thin branches or sprays of bamboo are often employed, but for simple compositions, a portion of the round stem or tube is selected, with a few leaf-clad twigs attached. The top of such tubes are cut off either in a splayed or hotizontal manner according to the occasion. If used at wedding feasts, the cutting must be hidden by leaves, the sight of it being considered unlucky and suggestive of severed friendship. These cylinders of green bamboo require very special and careful treatment in order to preserve their verdure and vitality for any length of time. They must be cut in the cool of the morning or evening. In the early months of Summer, when the leaves are young, the stem is very succulent, and no special treatment is required, but to preserve the freshness at other seasons, small holes are drilled between the knots of the tube, into which water is blown with the mouth. Sometimes *sake*, the rice-wine of the country, is added to the water as a stimulant.

Mostly three, but sometimes five twigs of leaves are left on these cylindrical stems, arranged irregularly. The leaves of such branches are, in their natural state, crowded and confused, and they therefore require thinning out, the withered ones being removed, and the remainder disposed in double or triple groups. Three different combinations of leaves are approved, that of the *Fish-tail*, in which two of the lancet-shaped leaves spread out like the tail of a common fish; that of the *Gold-fish-tail*, in which a central leaf is added, giving the group the resemblance to the triple-finned tail of a gold fish; and that called the *Flying-geese* shape, consisting of three sloping leaves, suggestive of the outline of a wild goose in flight as seen from a distance.

A

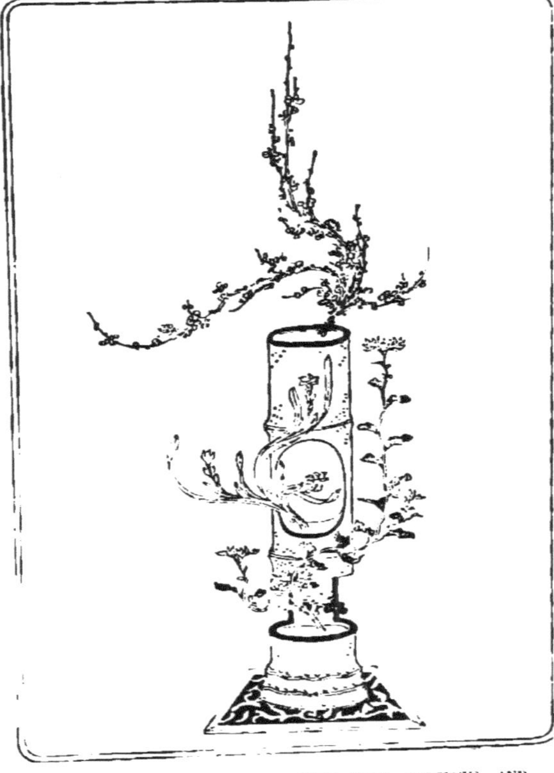

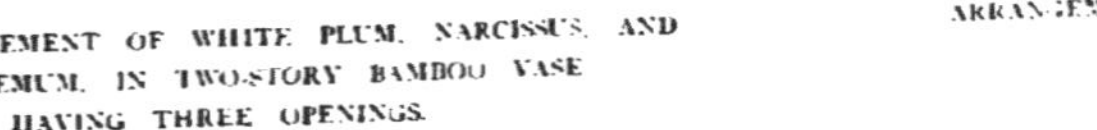

TRIPLE ARRANGEMENT OF WHITE PLUM, NARCISSUS, AND CHRYSANTHEMUM, IN TWO-STORY BAMBOO VASE HAVING THREE OPENINGS.

B

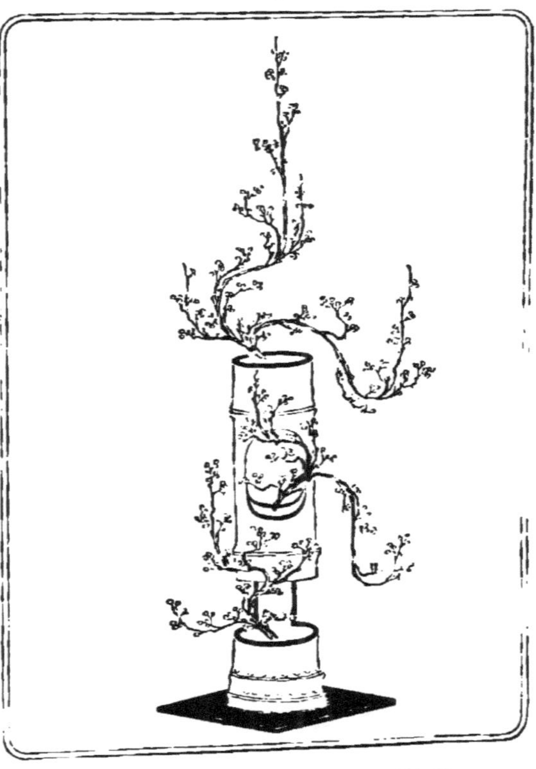

ARRANGEMENT OF *ILEX SIEBOLDI* (UME-MODOKI)

PLATE XLVIII.

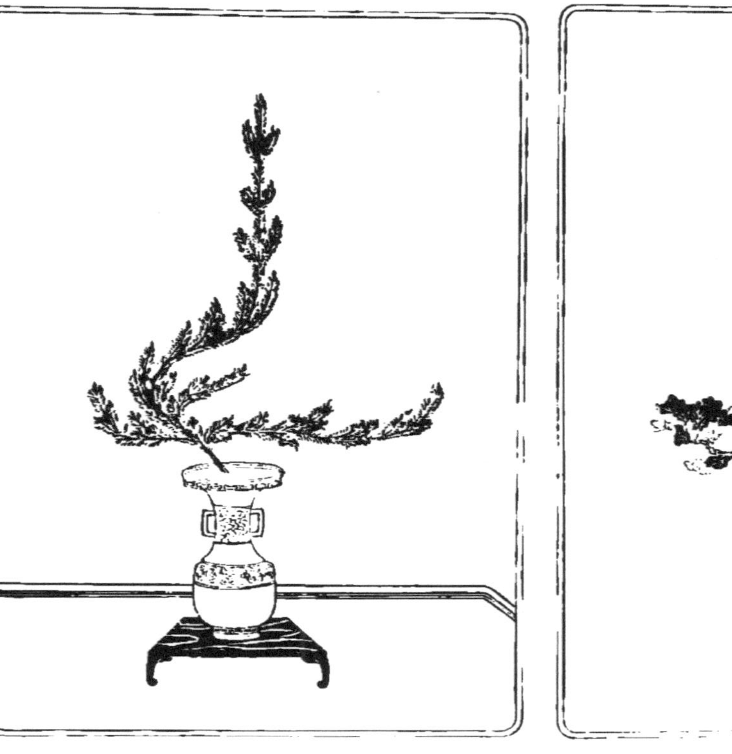

A

ARRANGEMENT OF SPECIES OF PINE (GOYO NO MATSU), IN STANDING BRONZE VASE

B

ARRANGEMENT OF CABBAGE-PLANT IN GLOBULAR VASE ON LEGGED STAND.

PLATE XLIX.

A

B

On some occasions, two tubes of green bamboo, one shorter than the other, are combined, being then called respectively the *male* and *female*. The top of the *female* or lower tube should be cut off horizontally, while that of the upper tube should have a splayed cutting. The longer tube has three knots or divisions, and two leaf-clad twigs which take the place of *Principal* and *Secondary* in the composition, and the shorter one only two knots and one spray of leaves, corresponding to the *Tertiary*. To designs placed in broad vessels a bamboo sprout (Take-no-ko) if in season, may be added as an auxiliary, and it should be fixed about two inches distant from the main stems, generally in a slanting position and close to an ornamental stone.

Numerous vases made of bamboo cylinders have been described in an earlier part of this work. Similar vessels are often made of green bamboo, the leaves being left on and forming part of the floral composition arranged in such vases. The same care is required to keep such receptacles fresh and green as is applied to ordinary bamboo tubes, and while the upper and lower portions are cut out to form mouths for the purpose of receiving other plants, the intermediate space between two knots is bored and sprinkled internally with water and *sake*. In Plate XLIX. at A, may be seen an arrangement of bamboo in a sand-basin, together with the *Nuphar japonicum*, and at B, a design with thin stems of bamboo alone.

COMBINATION OF PINE, BAMBOO, AND PLUM.

Separate notice is given to the combined arrangement of pine, bamboo, and plum branches, on account of the importance attached to this triple alliance in the art under consideration. This combination, called by the Japanese *Sho-chiku-bai*, is used at important celebrations in token of congratulation and well-wishing. It is specially employed at the New Year, and, if the plum blossom be in season, at wedding feasts. Sometimes the composition is in a single vessel, and sometimes three vases are used each containing one of the three growths. If all three be placed in one vase, the pine takes the position of the *Principal*, the bamboo of the *Secondary*, and the plum of the *Tertiary*; and if arranged separately, the pine occupies the central position, the bamboo is placed on the left, and the plum on the right. The plum branch should have its base tied with a coloured silken cord when used at weddings. A bamboo vase is not considered suitable for such triple arrangements, but a vessel of green bamboo may be employed with the pine and plum only, thus including the vase itself in the triple combination. An illustration of this is shown in Plate L., and another example of the *Sho-chiku-bai* may be seen in the same Plate, at B.

ARRANGEMENT OF WILLOW BRANCHES.

The willow is employed for floral designs on important occasions, from the fifteenth day of the eighth month to the first day of the third month. Owing to the facility with which its branches can be bent, it is considered the easiest of all *trees* to arrange after the Japanese manner, and generally forms the subject of first lessons in the Art. Among the many varieties of the willow found in Japan, those most used for floral compositions are the *Salix purpurea*, or ordinary river-side willow, and the *Salix babylonica* or weeping willow. When branches of the weeping willow are employed, care must be taken to avoid an arrangement in which lines droop on both sides of the same composition, such designs being only permitted at the celebration of death anniversaries. It is a common practice to tie the long trailing stems of the willow into a loose loop or ring of about four inches in diameter, generally three stems being so united, and the extremities hanging down to different lengths. The originator of the Enshiu School is said to have invented this style of composition, as he found it otherwise difficult when arranging willow branches in standing vessels, to prevent the long shoots from trailing on the floor. Another version attributes the practice of looping the willow to an old Chinese custom prevailing at farewell meetings, on which occasions it signified *tying-up* until the return of the departing guest: hence it is said to be a style of arrangement specially suited for farewell gatherings.

The custom became afterwards applied even to hanging compositions, which are generally preferred for this tree, the idea being that, because the willow grows near water, its branches should hang over laterally as if drooping over a stream. Even in chambers of handsome dimensions, for which standing flower arrangements are mostly selected, the willow is frequently placed in a vase hooked to the pillar of the recess, or in boat-shaped vessels suspended from the cross-beams. A design of a willow branch, in combination with a bunch of narcissus, placed in a hooked basket, may be seen in one of the text illustrations, Fig. 8.

The camellia is the favourite flower for introducing with the willow, a combination at variance with the general rule that two *trees* ought not to be employed together in double compositions. At the season when the leaf buds of the willow present a reddish appearance, the white camellia is chosen, but when these buds assume a greyish white colour the red camellia is considered more suitable. In Plate LII.a, is shown an arrangement of willow and camellia in a suspended bronze vessel of crescent

PLATE LI

shape, and in Plate LI. are illustrated two compositions, one consisting of the willow combined with the *Camellia sasanqua*, and another of the willow and narcissus. If both are arranged in a two-staged bamboo vase, the willow should be above, and the camellia below. The narcissus and winter chrysanthemum are also occasionally used in combination with the willow.

Fig. 26

Three fancy styles of composition are recognized for the branches of this tree, namely, the *willow in fair weather*, the *willow in wind*, and the *willow in snow*. For the fair-weather style the Spring willow is used, and the branches are spread as if just kept apart by the breeze; when the willow in wind is expressed, the branches are given a curved sweep, as if blown back by a strong wind, (see Fig. 26); and when the willow in snow is suggested, the stems are made to hang straight and heavily as if weighted with snow.

ARRANGEMENT OF WISTARIA FLOWERS.

The wistaria, as the first flower of Summer and the most important of blossoming creepers, holds high rank, though its colour renders it unsuitable for certain occasions,—purple being associated with mourning. The species with white blossom is seldom used in floral designs. From its nature this plant is specially adapted to suspended arrangements, though it is also used in standing vessels; in the latter case, some sort of frame is generally necessary for its support. The blossom-clusters are preferred before they are in full bloom, having a majority of buds and half-open flowers, with only three or four fully open.

Tubs and sand basins, such as water plants are placed in, are the only kinds of standing vessels used for the wistaria, which, it must be remembered, grows in parks and gardens on trellises overhanging a lake or stream. The idea of such water-scenery is conveyed by means of these broad, flat vessels, and stones are often introduced to add to the impression of landscape in miniature.

An arrangement with ornamental stones is shown in the accompanying woodcut, (Fig. 27). Such stones are generally three in number, with three distinct characters, distinguished by the names Heaven, Earth, and Mankind. The principal stone is high and pointed, like a mountain in shape; the third stone is flat and horizontal in character; and the second stone is of intermediary form. Other stones of minor importance are often added, but not so as to detract from the three essential ones, and the complete number should always be an odd one. The stem of the wistaria should spring from behind the principal stone. Water plants like the iris and *Nuphar japonicum*, and land plants such as the *Calendula officinalis*, Bijinso, and *Aspidistra lurida*, are sometimes combined with the wistaria in shallow vessels. If land plants are introduced, no stones should be employed, these being only resorted to when water scenery is expressed.

Fig. 27

A composition with wistaria, irises, and *Nuphar japonicum* would be arranged as follows: The wistaria should be placed to the left of the vessel, at the side of a high stone,—as if it were on the slope of a hill with its branches stretching over and its flowers reflected in an adjoining lake or stream. It should have five, seven, or nine clusters of blossoms. Then, at a distance of about six inches, a group of the leaves of the *Nuphar japonicum* is placed; and about four and a half inches from this plant are arranged the irises in a simple composition of three or five leaves, one of them curling over and dipping into the water. The remaining stones should then be distributed in different parts of the vessel, but so as not to interfere with the water plants, which must be placed in the free water-space.

The nature of the wistaria-creeper prevents its stems being arranged in distinct lines in accordance with the general methods of floral design, but the rules for *Principal*, *Secondary*, and *Tertiary*, are applied to the grouping of its blossom-clusters. In hanging arrangements the *Streamer* assumes great importance.

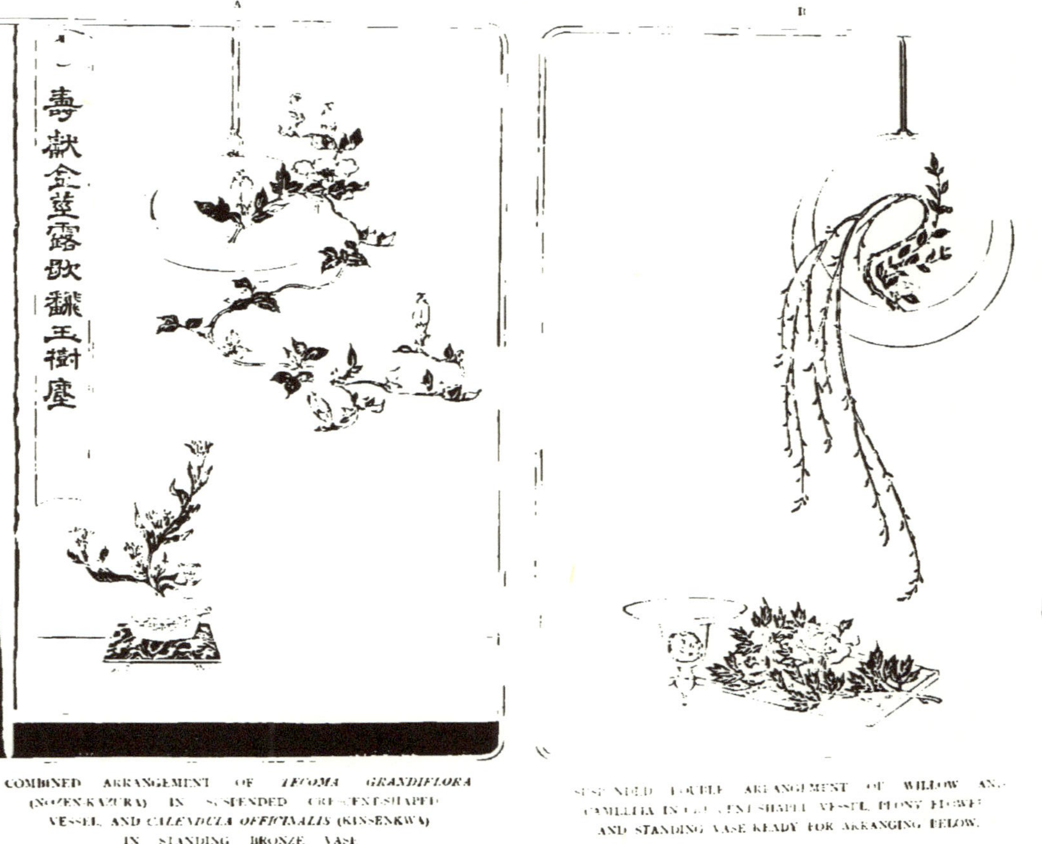

A: COMBINED ARRANGEMENT OF *TECOMA GRANDIFLORA* (NOZEN-KAZURA) IN SUSPENDED CRESCENT-SHAPED VESSEL, AND *CALENDULA OFFICINALIS* (KINSENKWA) IN STANDING BRONZE VASE

B: SUSPENDED DOUBLE ARRANGEMENT OF WILLOW AND CAMELLIA IN CRESCENT-SHAPED VESSEL, PEONY FLOWER AND STANDING VASE READY FOR ARRANGING BELOW.

PLATE LIII.

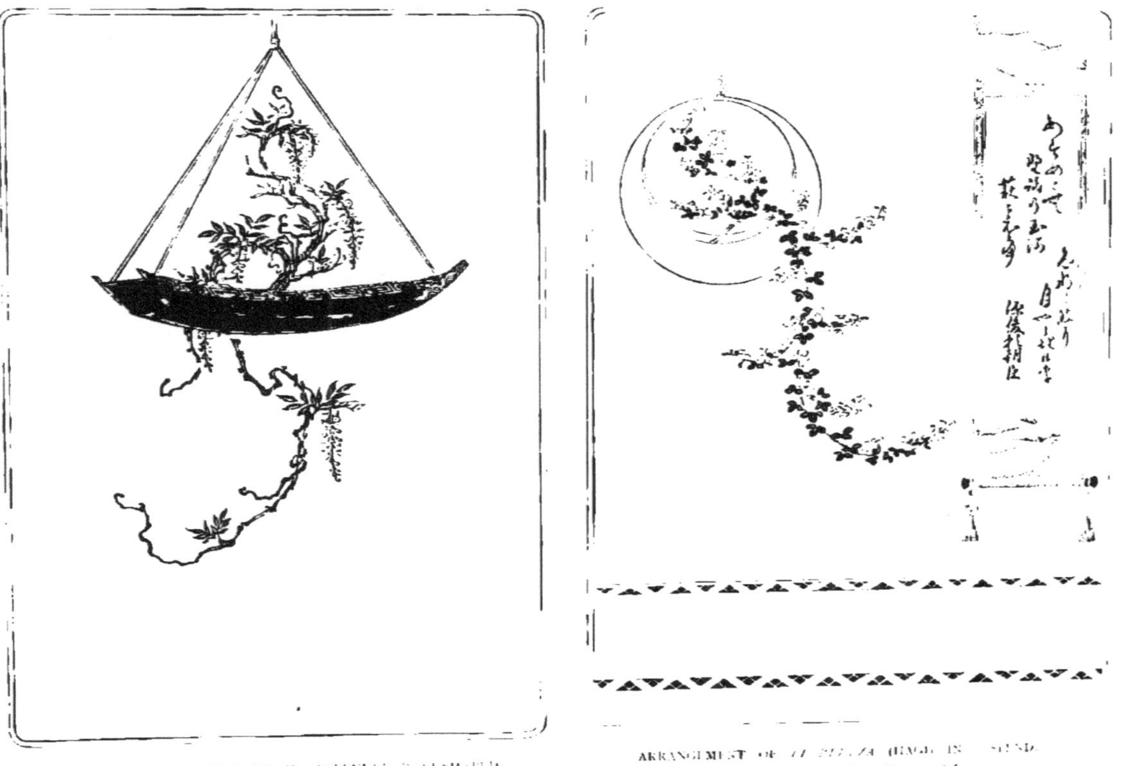

A

ARRANGEMENT OF WISTARIA IN SUSPENDED BOAT-SHAPED VESSEL OF BRONZE

ARRANGEMENT OF *[illegible]* (HAGI) IN [illegible] SUSPENDED CRESCENT-SHA[illegible] IN [illegible] HANGING MANUSCRIPT

Examples of suspended compositions may be seen in Plate LIII.A., where the flower is placed in a bronze boat-shape vessel, and in Fig. 16, page 76, where it is arranged in an inverted bell twining round a branch of pine.

ARRANGEMENT OF IRISES.

Several kinds of irises are employed for floral arrangements, the most important being two varieties of *Iris lævigata* (Kakitsubata and Hanashobu), the *Iris sibirica* (Ayame), the *Iris tectorum* (Ippatsu), and the *Iris japonica* (Shaga). Of these, the *Iris lævigata Fisch. var. Kæmpferi*, which the Japanese call Kakitsubata, is the favourite. This plant, though belonging particularly to the early Summer, lasts through several seasons, displaying a special character of growth at these different periods. In Spring, the leaves are stiff and straight, and the flower stems are short; in Summer, the leaves are more full and spreading and there is much spirit in the flowers; and in Autumn the flower stems are long and the leaves bent and curled. These peculiar characteristics of the plant at different seasons must be shown in the floral arrangements. The leaves of the iris, like those of other water plants, are considered the most important part of the composition, and they must be well selected, all withered ones being discarded. They are plucked as they grow in their sheaths, separated, and then artificially connected into groups of two or three, being attached with saliva. These sets of leaves are used to form the different lines of the design, names being given to each according to its position and function in the composition. The main leaves form the *Principal*, *Secondary*, and *Tertiary* lines of the arrangement, to these being added two or three flowers and other auxiliary leaves. Plates LV. and LIV. demonstrate the methods of taking to pieces the clumps of leaves, re-arranging them, and combining them, with the addition of flowers and buds, to form a simple composition.

The usual numerical proportions between leaves and flowers in an arrangement of irises, are :—one flower with three leaves, two flowers with seven leaves, two flowers with fifteen leaves; three flowers, with thirteen leaves; and five flowers with eleven leaves.

Taking, by way of example, a composition of two flowers and seven leaves, the following is the method of arranging. First, three leaves are placed in the position of the *Secondary* line. These three leaves are joined together, the two outer ones being long, and the central one short, as if just sprouting out from between the other two. Above this group of leaves a full-blown flower is placed, with one long leaf added, which occupies the position of the *Principal*. This is often called the *Cap-leaf*, as it crowns the whole.

Behind it a smaller leaf is added as an auxiliary, and at the side of this a flower bud is introduced. Next, a small leaf is placed in the position of the *Tertiary*, technically called the *Water-dipping-leaf*, because its tip curls over to the water of the vase; and combined with this is added another small leaf called the *Dew-supporting-leaf*, because its blade should be slightly concave above, with an upward tendency.

In making a composition of eleven leaves and five flowers, three leaves are put in for the *Secondary*, as before, and a flower is added. Behind this, and rather higher up, the *Cap-leaf* is placed, and with it is arranged the highest flower as *Principal*, with two other auxiliary leaves. Then another flower is added between the *Principal* and *Secondary* blossom with an additional leaf below it; and on the opposite side of the composition, a little lower than the flower of the *Secondary*, is placed another flower with two leaves adjoining. Still below this is added a bud, for the *Tertiary*, combined with two leaves,—the *Water-dipping-leaf* and the *Dew-supporting-leaf*. Of the flowers mentioned, two should be in full bloom, two partly open, and one in bud.

Reference has been made to flower arrangements in broad flat vessels in which trees or plants are disposed side by side in divided clumps. The distance between such clumps in the case of *tree* arrangements, is called the *Valley-space*, but in that of *water-plants* it is called the *Fish-travelling-distance*, because a lake or stream containing such plants is suggested, and the space left is just sufficient for the passage of large fish. Such interval should measure about one tenth of the height of the largest group of plants.

In a divided arrangement of irises in two clumps, one should be large and high, in the *Vertical-triangle* style, (see Fig. 1, page 47), and the other small and leaning, in the *Horizontal-triangle* style, (Fig. 2, page 48). The main group, placed to the right side of the basin, may have seven large leaves, a flower, and a bud; and the other group, six small curled leaves, and one bud.

The Japanese floral artist delights to suggest in his compositions some noted natural landscape. The most famous spot in Japan for displays of the iris, is Yatsuhashi, in the province of Mikawa, where there is a river with eight tributary streams crossed by as many bridges. An artifiical flower arrangement is sometimes made in suggestion of this natural view, by using a very large, shallow basin in which are placed white pebbles to represent the river and its eight branches, divided by black stones, filled in for the land.

PLATE LIV.

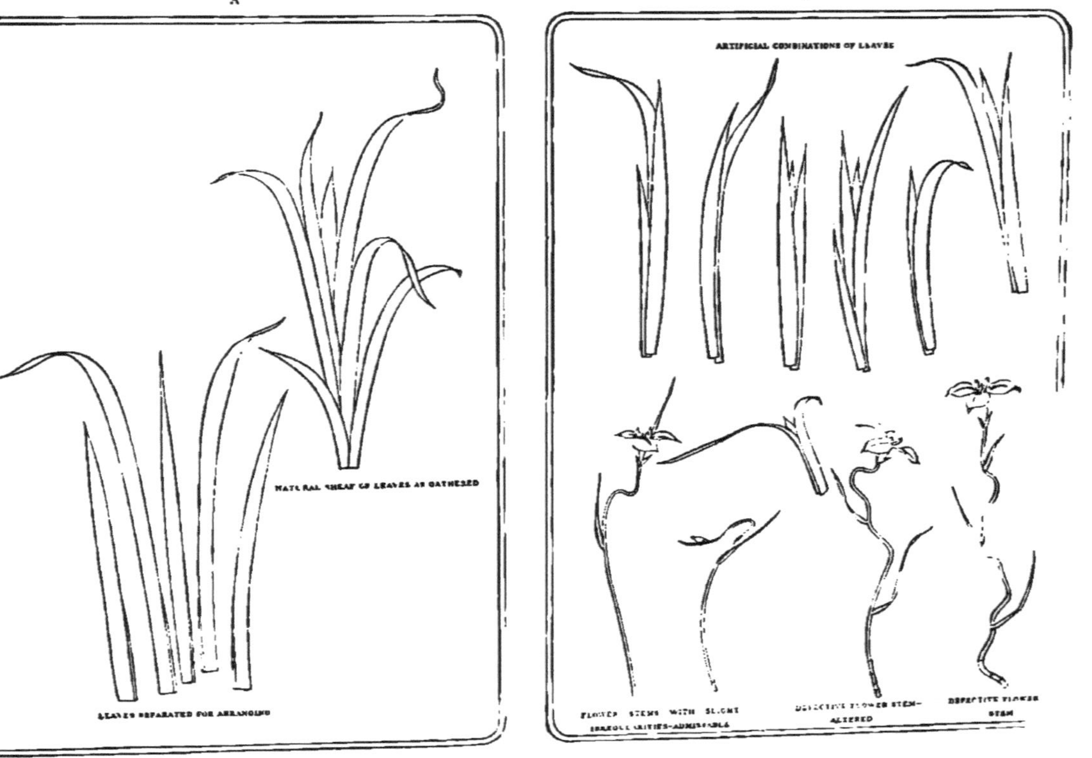
A
B
NATURAL SHEAF OF LEAVES AS GATHERED
LEAVES SEPARATED FOR ARRANGING
ARTIFICIAL COMBINATIONS OF LEAVES
FLOWER STEMS WITH SLIGHT IRREGULARITIES—ADMISSABLE
DEFECTIVE FLOWER STEM—ALTERED
DEFECTIVE FLOWER STEM

PLATE LV.

A

ARRANGEMENT WITH TWO FLOWERS

ARRANGEMENT WITH ONE FLOWER

ARRANGEMENT WITH THREE FLOWERS

B

The irises are then arranged amongst the black stones in divided groups. Each group is composed separately with a different quantity of leaves and a proportionate number of buds and flowers. Some bunches are disposed vertically, others in a more leaning style, and numerous pairs of young leaves are arranged between, to connect the whole. This arrangement is one example of a kind of flower composition almost resembling miniature gardening. The plants are held in their position by the sand and pebbles, and no other visible fasteners should be employed.

The *Iris sibirica* is arranged in a somewhat different manner, the flowers being placed higher than the leaves, whereas, with the *Iris lævigata*, the blades rise above the blossoms. Another peculiarity is that the leaves are arranged in regular gradations and not in indented triplets.

The Iris may be seen in Plate LVII., at A, arranged in a large basket together with a branch of fir, and at B, it is shewn as an independent composition in a small bamboo vase. One of the subjects in Plate XXVI.A. is that of flags placed in a *Well-frame* vessel, and held by a metal *Crab-fastener*. In Plate LVII. a large design of irises is illustrated, forming one side of a paired floral arrangement, the *Aster tartaricus* being placed in another vase opposite to it. The iris is often arranged with other water-plants, such as the *Nuphar japonicum* and *Rhodea japonica*, in a broad vessel. It is also to be seen combined with trees, such as the wistaria, maple, and *Nandina domestica*.

ARRANGEMENT OF PEONIES.

Two distinct varieties of this plant are employed for floral designs, namely, the *Pæonia moutan* and the *Pæonia albiflora*. The former produces blossoms of immense size, and is generally arranged in wide-mouthed bronze vases or antique flower-baskets, no other flowers being permitted in conjunction. The *Pæonia albiflora*,—a smaller species which goes by the name of *Shakuyaku*,—may be used in combination with branches of flowering *trees*, but should not be joined with other *plants*. The *Pæonia moutan*, or large peony, is one of the three flowers to which are attributed royal rank, the other two being the cherry and the lotus. It must therefore be used alone, and placed in the position of honour in a chamber,—that is, on the dais of the principal recess. It should never be arranged on a shelf or in any secondary place, and all other flowers must be excluded from the room which it is used to decorate. This rule is professedly followed out of respect for the peony as queen of all flowering plants, but its luxurious character really renders such practice consistent with purely artistic considerations, the addition of other material being calculated to produce a surfeit of richness and elaboration. The employment of the cherry blossom is also subjected to the same restrictions.

It is customary in arranging the large peony to introduce between the principal stems one or two black withered twigs, which, by contrast, enhance the appearance of the leaves and flowers. As the blossoms are massive and heavy in character, they should be sparingly introduced between the leaves, rarely more than one full-blown blossom being used: the remainder consists of buds and partially opened flowers. For the ordinary composition in three lines, five or seven blossoms may be employed, one for the *Principal*, one for the *Secondary*, one for the *Tertiary*, and the remainder distributed in intermediary positions. The leaves are carefully disposed so as to give *support* to the flowers, and in some cases are made to cross in front of them. The leaves surrounding the *Secondary* blossom should be large and closely grouped. They are technically called the *Lion-hiding* leaves, because the presence of the lion, which is associated with peonies in decorative art motives,—is supposed to be suggested behind the thick foliage. A flower bud should be used for the *Tertiary*, with the addition of strong young leaves.

The *Paeonia albiflora*, or small species of peony, is not held in so much honour as the larger kind, and is often combined with other plants. In the case of this plant, half opened blossoms are preferred for the *Principal*, full blossoms for the *Secondary*, and buds for the *Tertiary*. The peony is sometimes arranged in wide basins divided into separate groups.

Illustrations of this flower, arranged in different kinds of vases, may be seen in Plate LVIII, at A. and at B.

ARRANGEMENT OF LOTUS FLOWERS.

As has been stated in an early part of this work, the lotus plant is associated with the spirits of the dead, and is therefore considered out of place as a decoration at festive gatherings. It, however, holds high rank in the Floral Art, being regarded as king of the flowers of India the source of Buddhism. When employed on the dais it must be given precedence of all other flowers, and it would be considered a gross violation of taste to combine with it the peony, which the Japanese regard as the royal flower of China. In the case of the lotus plant, the leaves

PLATE LVI.

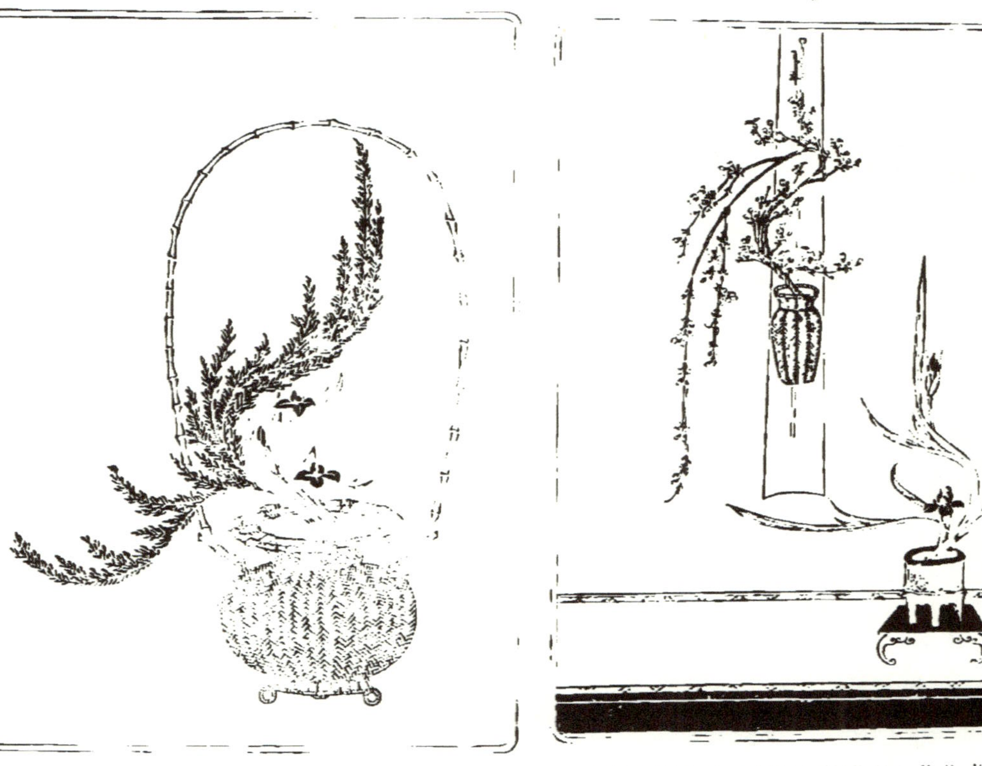

A

ARRANGEMENT OF FIR AND IRIS IN STANDING FLOWER-BASKET WITH ARCHED TOP HANDLE.

B

ARRANGEMENT OF WEEPING-CHERRY IN A [illegible] BASKET AGAINST PILLAR TABLET [illegible] WITH IRIS IN STANDING BASIN.

PLATE LVII

PLATE LVIII

A

PLATE IX

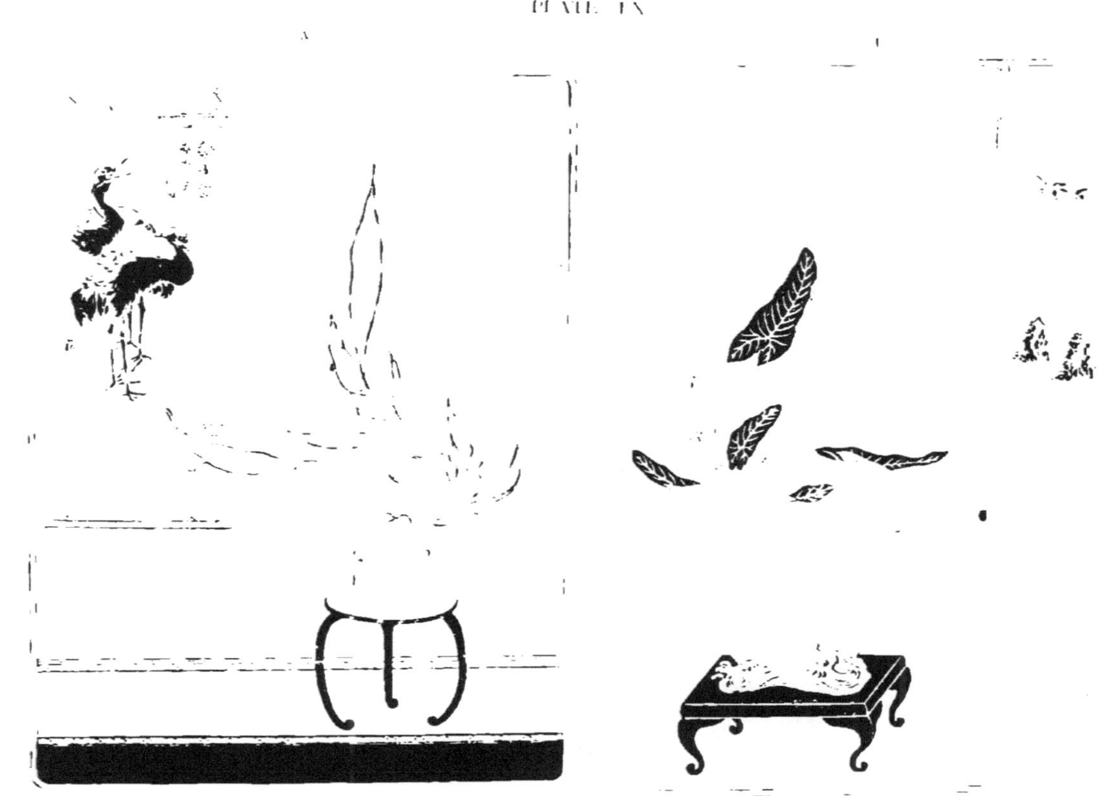

PLATE LXI.

play the most important part in the composition, this being a rule which applies to nearly all water-plants. The lotus leaves should be selected to express the idea of the three Buddhist divisions of time—present, past, and future. *Past time* is represented by a partly decayed or worm-eaten leaf; *Present time*, by a handsome open leaf,—often called the *Mirror-leaf*, on account of its resemblance in shape to that of a Japanese mirror; and *future time*, by a curled leaf not fully open.

This plant should be arranged in a wide-mouthed vase, *Sand basin*, or *Tub*, in which vessels it is often united with other water-plants. As an example of such compound arrangements may be described a combination with the iris, *Nuphar japonicum* (Kohone), *Alisma plantago* (Omodake), and rush, with the addition of ornamental stones, so as to form a sort of miniature lake scene. A general rule applying to floral designs in broad basins, is that tall plants suggest shallow water, whilst those short in growth suggest deep water; and as the lotus flourishes in comparatively deep water, the stems of its leaves, in a compound design, are kept short. The idea of *Principal Secondary* and *Tertiary* is followed in the disposition and character of the leaves; for the *Principal*, a large round leaf being employed, for the *Secondary* a smaller leaf, and for the *Tertiary* a young and curled leaf. An open blossom is introduced between the *Principal* and the *Secondary*, and a flower bud, kept low in position, is placed between the *Principal* and the *Tertiary*. The irises are next arranged at a distance of about two and a half inches from the lotuses, an open flower being used for the *Principal*, a half open flower with three leaves for the *Secondary*, and below, in the place of the *Tertiary*, two or three more small leaves. The *Nuphar japonicum* is next placed in position, its *Principal* consisting of a large oval leaf of interesting shape, and below it a half open flower occupying the place of the *Secondary*, whilst for the *Tertiary*, a young and only partly opened leaf is used. The rushes are then arranged in five or seven bunches, and the *Alisma plantago* is afterwards introduced, separated from the rushes by a short space, with two leaves having a bunch of flowers between them, the composition being similar in style to that of the *Nuphar japonicum*, but bent in an opposite direction. Of the five plants in this combination the lotuses and rushes are disposed in the *Vertical-triangle* style, and the others in the *Horizontal-triangle* style. After they are all arranged in position, the stones are added to connect the whole group together.

ARRANGEMENT OF CHRYSANTHEMUMS.

Floral compositions with chrysanthemums are considered very difficult to arrang skilfully. Of this flower there are many kinds, some of larg and some of small blossom, the latter presenting least difficult in treatment. There are varieties peculiar both to Summ and Autumn, and also a later species exists called the Winte chrysanthemum. Characteristics of growth at the differe seasons are faithfully imitated in floral compositions. In Fig. 28, are shown there separate sprays of the small Summe chrysanthemum, and the same combined to make a flor design. It will be observed that the lines of this compos tion are strong and vigorous, corresponding with the charac ter of Summer growth. In contrast to this, Fig. 29 shows an arrangement of Autumn chrysanthemums, which is altogether more mature in style, in accordance with the character of Autumn vegetation.

Fig. 28.

This plant requires great care in order to preserve freshness in the leaves. It should be cut after sunset, and the ends of the cuttings should be burnt to charcoal, and placed for a considerable time in water before wedging, so as to prolong their vitality. The stems are brittle, and need great skill in bending to the desired curves, very refractory por-

Fig. 20.

tions being softened by heat. The blossoms of the small chrysanthemum, as shown in Figs. 28 and 31, are massed in groups, generally consisting of an uneven number of flowers. Figures 29 and 30 illustrate the methods of grouping together small chrysanthemums as auxiliary to a branch of some tree, in a mixed composition. In Fig. 29, three sprays are used, and in Fig. 30, five are employed. With the species of large flower, the blossoms are isolated and of limited number, each stem bearing one, two, or at most three flowers. There are said to be seven faults which must be guarded against in the disposition of such large flowers. A blossom must not be turned away so as to present its back in the composition; nor must it turn its full disc to view; flowers must not have stems of the same length so as to be exactly in a line; nor must three blossoms be arranged in a triangular form; the disposal of blossoms in regular steps is objected to; *colour-sandwiching*,—as previously described,—is another fault to be guarded against; a large open blossom should not be put at the lower part of the composition; and flowers should not be hidden by leaves. To correct a tendency to the above faults small bamboo skewers carefully hidden are employed. The illustration Fig. 32 shows the method followed in arranging plants having large blossoms or large clusters of flowers. The flower shown

Fig. 30.

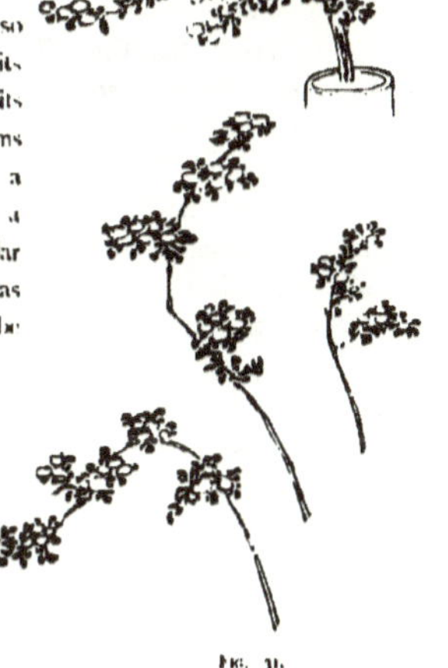

Fig. 31.

is the patrinia. There are seven principal flower clusters, which are essential, marked A, B, C, D, E, F, G, corresponding to the masses of a seven-lined composition,—*Principal, Secondary, Support, Tertiary, Sub-principal, Side*, and *Trunk*; and the additional flowers d, e, f, g, and others, are merely auxiliary and added at discretion, to avoid bareness.

In addition to the special form of arrangement suggested by the peculiarities of growth at different seasons, three general styles of composition are recognized for floral designs; namely, the *finished* style, the *intermediary* style, and the *rough* style. These distinctions correspond with, and are described by, the same terms as those applied to the square, and grass or running Chinese ideographs, and refer to different degrees of elaboration or *sketchiness*. In Plate XLVII.A, are shown chrysanthemums arranged in a *rough* style in combination with other flowers. Plate LXII. exhibits two compositions of chrysanthemums in a more *finished* style, one representing these plants disposed in five lines, with seventeen blossoms, in a fancy bronze vase, and the other showing a triple design in a bamboo vase of three mouths, with fourteen blossoms in all. A *rough* arrangement of these flowers in a special kind of basket is illustrated in Fig. 9, page 61.

Fig. 32

ARRANGEMENT OF *NUPHAR JAPONICUM*.

The *Nuphar japonicum*, on account of its beautiful oval leaves, is a favourite water-plant for floral designs. It is arranged in *Sand-basins* and *Tubs*, the *Horse-tub*

PLATE LXII

A

B

being by preference selected, and the *Horse's-bit*, or some other fancy fastener of metal employed. The largest leaves of this plant hang horizontally on their stems, and are gracefully curled at their edges like those of the lotus. The stems are generally arranged as long as possible to convey the impression of a plant which grows in shallow water. This rule becomes specially important when the nuphar is combined in a single vessel with other water-plants which flourish in less shallow pools; such deep-water plants being purposely kept as short as possible and placed more centrally so as to suggest the idea of greater depth. Though the flowers of this plant are small in comparison with the leaves, their position in a design is much studied, and the leaves are arranged in reference to them. Seldom more than two or three are introduced into a single composition, and these, consisting of a bud and one or two half open blossoms, are grouped between the *Principal* and *Tertiary* leaves.

For the *Principal*, a large and slightly curled leaf is employed, placed higher than the rest, and extending to the left of the design; for the *Secondary*, a somewhat smaller leaf is fixed leaning to the right; for the *Tertiary*, a small rolled leaf, technically called the *Horn-leaf*,—is added below the *Principal*, and beneath this is a young bent leaf called the *Water-touching-leaf*, which should be only about an inch above the water surface.

The slender stems of this plant when arranged as above show a deficiency in width, and it is usual in broad water-vessels to introduce a secondary clump at the side, removed from the main composition by the space of a few inches. This may consist of a single flower and one small leaf, kept very short as if just sprouting from the water.

Double-well-buckets are often chosen for arrangements of the *Nuphar japonicum*. In Plate LIX.A. may be seen such a composition in the lower of a pair of suspended *Well-buckets*, the clematis being introduced into the upper one, in a hanging style, with a long *Streamer*. Another illustration, (Plate LX.B.), shows the same water-plant disposed with seven leaves and two flowers, and in Plate XLIX.A. it is combined in a large *Sand-basin* with a high group of bamboos.

ARRANGEMENT OF *KERRIA JAPONICA.*

This modest plant, with its long sprays of yellow blossoms, grows in abundance near the banks of rivers, and is specially associated with the scenery of the Tama-gawa, at Ide, near Kioto. The accompanying illustration, Fig. 33, represents a composition made with the Kerria in a broad *Sand-basin*, intended to suggest such river-scenery. The stems are held by long cylinders of bamboo basket-work made in imitation of the *Ja-kago*, or huge baskets holding boulders which are used to break the current of rapid streams. Stones are added to complete the idea of the natural view.

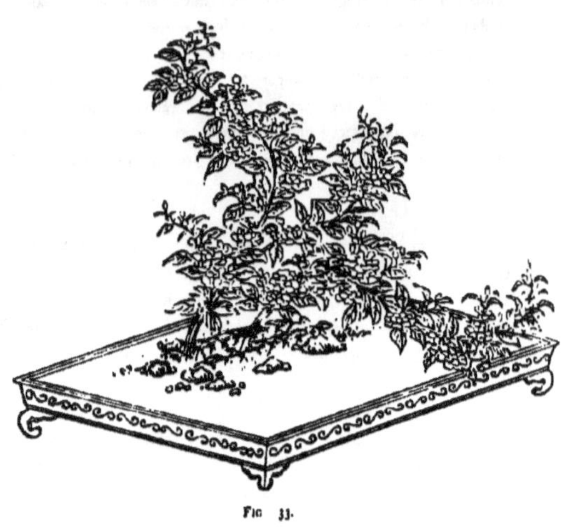

Fig. 33.

The *Kerria japonica* is also frequently arranged in suspended receptacles and *Boats*, the stems having a leaning character given them with the object of preserving the notion of flowers hanging over and reflected into a stream. The popular fancy that this plant bears neither seed nor fruit prohibits its employment at wedding ceremonies.

Plate XLII. illustrates a defective and a corrected arrangement of the Kerria in standing vases, and Plates XXXVI. and LXIII. show the same flower disposed in a *Horse-tub* with the *Horse's-bit* used as a fastener.

ARRANGEMENT OF NARCISSUS FLOWERS.

The narcissus takes an important place in floral arrangements during the Winter and Spring months. Available at a time of year when other blossoming plants are scarce, sweet in scent, and of golden colour, it is considerably prized by the flower-artist. Varieties of red, white, pale blue, and double blossom, exist, but with the exception of the white, which is sometimes used on ceremonial occasions, these rare specimens are seldom seen in floral compositions. The blade-shaped leaves of this plant receive chief attention, and are used in much greater number than the flowers; they are removed from the sheath, and are separately pressed and rubbed on a board to take out the excessive twist which the heat of the sun gives them in their natural state, a single curl alone being permissible for leaves employed in compositions. The younger leaves are easily smoothed and straightened with the fingers, but the large ones are refractory and require much labour and patience. Thus altered they are attached together in pairs consisting of one long and one short leaf, with the longer blade in front, or sometimes in triplets, and these bunches are employed for the different lines of the arrangement, with the flowers placed between, and always lower than the leaves. Fig. 34 shows first a narcissus plant pulled to pieces and its sheath removed, then the method of smoothing the leaves with the fingers to take out the curl, and afterwards the way in which the manipulated leaves are connected in pairs and fours by mean of saliva and the replaced sheath. Fig. 35 shows different artificial combinations of these leaves of which the highly curled are only sparingly introduced into compositions; and in Fig. 36, is illustrated a small bunch of leaves with flowers tied with paper before immersing in water, also a trilineal completed arrangement consisting of combined groups of leaves and flowers. In Fig. 37 different kinds of twists given to leaves are shown, those in C and D are considered affected and exaggerated, and those in A and B are preferred. Leaves which curl are chiefly used in hanging compositions. The lengths of the flower stems are made to vary according to the particular season, being shorter in Winter than in Spring arrangements. If the blossoms droop they are carefully straightened and held upright by small

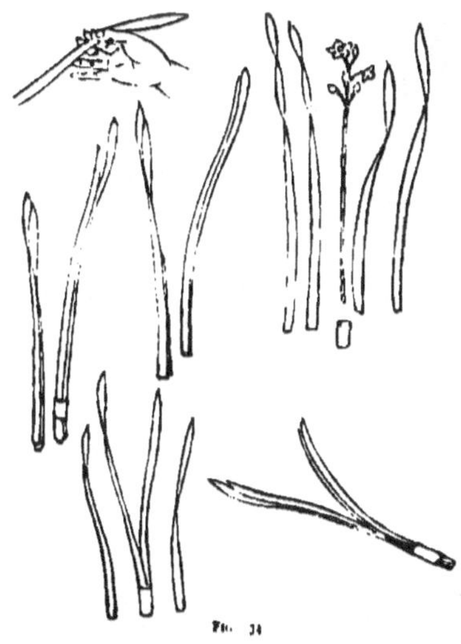

FIG. 34

wooden spikes. A simple five-lined composition with narcissus is shown in Plate LXIV.a a double arrangement of willow a narcissus is illustrated in Plate LI.A., a in Plate XLVII.A. may be seen a tri composition in a high bamboo vase, which the narcissus occupies the cent position, with plum branches above a chrysanthemums below.

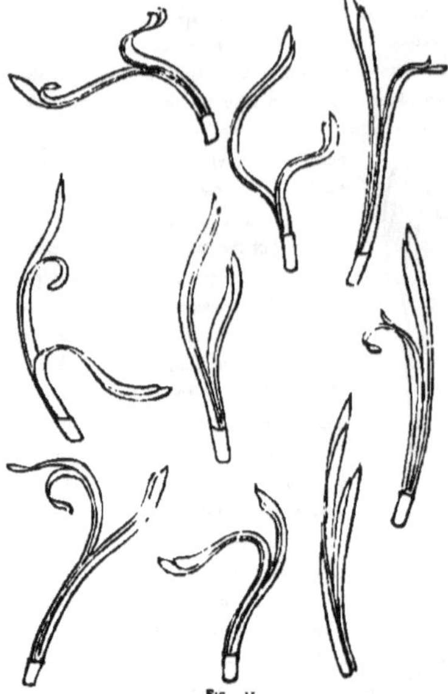

Fig. 35.

ARRANGEMENT OF CHERRY BLOSSOMS.

The cherry blossom, being garded as the king of indigenous flowe has precedence of all others for floral rangements. It is considered preferable to use branches of cherry blossom alone, and they are but rarely combined with other flowers. Sometimes, however, combinations with pine branches, with rape blossoms, (*Brassica chinensis*), and with irises, are to be seen, though they are hardly considered orthodox designs.

Fig. 36.

It is said that, in making arrangements with cherry blossoms, the scissors must not be used, Though this rule is not strictly followed, the branches should be trimmed and bent as little as possible, and the composition must partake more of a natural than an artificial character. In accordance with the distaste for over-exuberance in floral designs, the cherry of small and single blossom is selected in preference to the richer specimens of double blossom.

A composition made with cherry flowers should have a somewhat sparse arrangement of buds at the top of the central line, half-open blossoms in the middle, and a few in full bloom below. Several fallen petals should be placed in the water of the vase, and any blossoms which have dropped may be left remaining on the flower stand. From the above rules it will be observed that the prevailing idea is to represent the cherry-tree as nearly as possible in its natural state in which the lower branches are the first to burst into bloom. The scenery of picturesque spots famous for the wild cherry-tree, supplies the motive for certain compositions. In Fig. 13, page 68, is shown a simple bunch of blossoms arranged in a small flower-basket placed on a miniature raft, intended to suggest the flower-laden boats which float down the rivers at Yoshino and Arashiyama during the cherry season. In Plate LVI. at B, is illustrated a composition made with sprays of the weeping cherry placed in a hooking flower-basket.

Fig. 17

ARRANGEMENT OF PEACH BLOSSOMS.

The peach blossom can hardly be said to receive attention in Japan in proportion to its great beauty and richness. The peach trees in flower form an attractive feature of the Spring landscape, but their gathered branches, though often employed in floral designs, are somewhat too exuberant for the taste of the flower artist. The peach is used chiefly in flower compositions during the girls' festival of the third month. The cuttings are then considerably thinned out, most of the full-blown flowers being removed, and only a small number of half-open blossoms and buds left distributed in masses amongst the foliage. The flower of double blossom is never employed, as it is said to have poisonous properties.

In Plate LXV. at A, is shown an arrangement of peach blossoms in a standing flower-basket, together with the *Rosa indica*, and in the same illustration at B, the peach is arranged alone in a hooking basket.

ARRANGEMENT OF CAMELLIAS.

The red and white camellia, both single, and double, are much used for Spring arrangements, though generally in combination with other trees or flowers. Their employment in conjunction with the willow has been mentioned in discussing willow compositions. There is a prejudice against the camellia on account of a peculiarity of its blossom, which does not fall to pieces petal by petal like other flowers, but drops off bodily, suggestive, it is said, of a head struck off with the sword. On the other hand, it is held in considerable estimation as a tree indigenous to Japan which has been much used for various purposes from ancient times. It is also highly esteemed as an evergreen In floral designs the disposition of the glossy oval leaves receives the greatest attention; two leaves at least must be appended to each bud or blossom introduced into the design.

Sometimes these two leaves point upwards, extending above the flowers; sometimes they are bent over more horizontally, so that the whole of the blossom appears above them. The white camellia, if in season, is much used at *Coming-of-age* celebrations. The favourite kind of composition is one of five blossoms and fifteen leaves, each flower or bud being surrounded by three curling leaves.

Examples of designs in which the camellia is introduced may be seen in Plate LII.B, where it is combined with the willow in a suspended bronze vase of crescent shape; in Plate XXV.A, it is connected with the peach and narcissus. In Plate LI.B, a particular kind called the *Camellia sasanqua* is shown, arranged with the willow. Bamboo vases are specially suited for camellias, those of the *Lion's-mouth* shape being selected by preference.

ARRANGEMENT OF CONVOLVULI.

The limp and delicate stems of creeping plants like the convolvulus and morning glory, are difficult to arrange according to the lineal rules of the Flower Art. Thin bamboo rods, withered stumps, or twigs, are used as a basis in forming a composition, and the design consists in a judicious balancing of leaves, flowers, and buds, in which can be detected the radical elements of *Principal*, *Secondary*, and *Tertiary*. For the lower portion of an arrangement, flower buds are preferred to open blossoms. Rikiu is said to have originated a composition consisting simply of one flower and one leaf of the con-

PLATE LXIII.

A

B

ARRANGEMENT OF *NANDINA DOMESTICA* (NANTEN) IN FANCY BRONZE VASE ON STAND

ARRANGEMENT OF *KERRIA JAPONICA* (YAMABUKI) IN "HORSE TUB," STEMS BEING HELD BY "HORSE-BIT" FASTENER

PLATE LXIV.

PLATE LXV

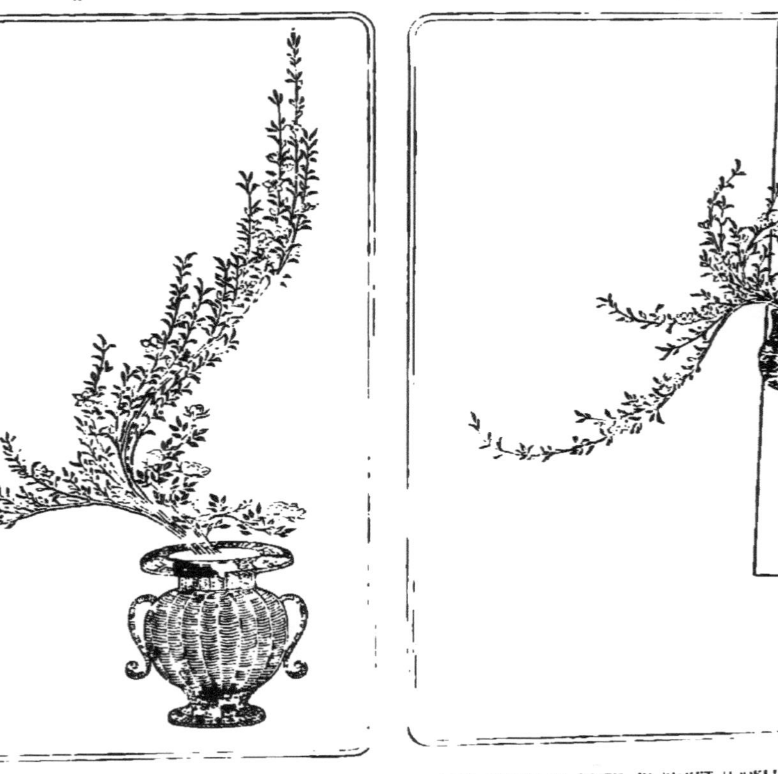

A. ARRANGEMENT PEACH AND *ROSA INDICA* (CHOSHUN) IN STANDING BASKET WITH SIDE HANDLES

B. ARRANGEMENT OF PEACH, IN BASKET HOOKED ON TO TABLET

volvulus, which afterwards became a favourite decoration for *Tea-rooms*. Upon being asked why he adopted so modest a design, he replied that, as it was impossible to rival nature in her grouping, artificial arrangements should be as unassuming as possible; even a single flower with one leaf being sufficient, he maintained, to call for admiration. One is tempted to believe that the difficulty of arranging creeping plants according to the accepted standards of lineal balance, may have had more to do with such a departure than the experienced master was ready to admit. The founder of the Enshiu School invented a method of employing the convolvulus in a flower-basket, by winding the creeper round the long oval handle of the receptacle. Such a composition is illustrated in Fig. 6, page 58. A convolvulus of three blossoms, placed in a standing vase on a high table, is shown in Plate LXI.A.

ARRANGEMENT OF LESPEDEZA FLOWERS.

The lespedeza is the principal of the Seven Plants of Autumn, and is much used for floral designs at this season. On account of its numerous small oval leaves and tiny flowers, a full and crowded arrangement of sprays is generally resorted to. When placed in standing vases a trilineal composition is followed, each *line* consisting of a group of four or five stems richly loaded with leaves and blossoms. The lower or *Tertiary* line may be composed of a number of short stems only sparsely supplied with leaves and flower buds, and curved in such a manner as to suggest the profile of a wild boar's head. The wild boar is supposed to sleep beneath the lespedeza branches, and is associated in art with this Autumn plant. There is an early specimen called the Summer lespedeza, and this should be arranged in a thinner and more open manner than the Autumn plant, in order to preserve its special character of growth, which is less profuse than that of the later season. In hooked or suspended receptacles, the lespedeza is employed with a long *Streamer*, the other branches being kept short and disposed in a simple and quiet manner, Such an arrangement, placed in a crescent-shaped vase, is illustrated in Plate LIII.B. The lespedeza is often arranged in suspended bronze boats. It is also sometimes used with the morning glory, the *Eularia japonica*, *Valeriana villosa*, *Valeriana officinalis*, *Pueraria thunbergiana*, and the carnation; making together the combination called the Seven Plants of Autumn.

ARRANGEMENT OF THE *RHODEA JAPONICA*.

This water-plant is valued for its beautiful large leaves, which are arranged with the greatest care and precision, each important one receiving a special name according to its position or function in the design, as follows:—

The *Central-leaf*, which occupies the place of the *Principal*, or the middle point of the arrangement. The *Spring-leaf*, a young and curled leaf enveloping the central stem. The *Autumn-leaf*, placed in the background of the design, and having a faded or withered tip. The *Dew-supporting-leaf*, employed in front of the *Principal*, in Spring and Summer arrangements, and having a curl upwards. The *Dew-spilling-leaf*, used behind the *Principal* in Autumn and Winter designs, and having a curl downwards. The *Swallow's-mouth-leaves* consisting of two young leaves lapped together so as to present a double tip, and placed between the larger leaves of the composition. The *Frost-protecting-leaf*, which bends over some distance above the berries of the rhodea, as if to shelter them from snow or frost. The *Wind-protecting-leaf*, which also curves over the berries, somewhat lower in position, as if to screen them from wind. The *Berry-protecting-leaves*, being two or three leaves disposed below and around the berries and from between which they appear to spring: in some arrangements these consist of withered leaves. It is considered imperative that, in floral designs made with this plant, the leaves should exhibit a proper balance of front and back surfaces. The colour and gloss, to which great attention is given, are much improved by sprinkling the leaves with *sake*, and immersing their stems for several hours in the same liquid.

In disposing the leaves in a composition they are piled in a step-like manner in pairs and threes. In this way, for a seven-leaved design, first a pair of leaves are placed, then another pair, and lastly three together; the berries are always introduced low down, screened and partly hidden by the foliage.

Arrangements of six different styles are recognized for the *Rhodea japonica*, as follow:—

The *Rainy-season-arrangement*, suited for any period of the year, in which style both the *Dew-spilling* and *Dew-supporting* leaves are employed. The *Snow-time arrangement* in which the *Frost-protecting* and *Berry-protecting* leaves are introduced, all withered leaves are discarded, and the berries are kept as low as possible in position. The *Spring-arrangement*, suited only for the early months of the year, in which many young leaves are used. The *Summer arrangement*, distinguished by the introduction of a number of large leaves, with the addition of one withered leaf in front. The *Autumn-arrangement*, in which several withered leaves are placed and no young leaves allowed. The *Winter-arrangement*, consisting of a majority of withered leaves with the addition of the *Swallow's-mouth-leaves*.

ARRANGEMENT OF LEAF ORCHID.

The Chinese orchid, called *Baran*, is one of the principal subjects for leaf arrangement in Japanese floral design, and its treatment serves as a model for most compositions made with large-leaved plants. The flowers, being small and insignificant, are often omitted, but, when employed, they are attached by means of thin spikes of bamboo to raise them slightly in position. The *Baran* requires very careful treatment in order to preserve its freshness. It must be cut in the early morning or after sundown, and its leaves are then curled up, tied with string, and immersed in water for some hours before use. In very hot weather it is customary to suspend the cuttings for half a day in a deep well. To give a good colour and gloss to the leaves, *sake* is forced up their stems before immersing in water.

This plant is generally arranged in a water-basin, sand-basin, tub, or other wide-mouthed receptacle, with the addition of ornamental stones. In its natural state the leaf-orchid has always one long oval leaf growing centrally and higher than the others, and in floral compositions this is used as the *Principal*, and is called the *End-leaf*. The bottom leaf of a series is small, with its point arching over; it corresponds to the *Tertiary* in floral arrangements, and is called the *Finishing-leaf*. Another special leaf introduced into certain compositions is one curled up spirally, and called the *Spider's-leaf*, being copied from leaves which are curled by the spinning of insects. This form is artificially produced by heating. A ragged leaf called the *Decayed-leaf*, made by tearing and scraping, is occasionally added. All the leaves of a composition including the above, are arranged in positions corresponding with the radical lines of a floral design, each leaf counting separately in the combination. Thus, in a three-leaved composition, the *End-leaf* will be used as *Principal*, the *Finishing-leaf*, *Decayed-leaf* or *Spider's-leaf* as *Tertiary*, and an intermediary leaf will occupy the position of *Secondary*. In the same manner, for a larger composition of thirteen leaves, the *Finishing-leaf* is placed as *Principal* at the top and centre of the design; to the right and below are arranged four leaves called respectively *Secondary*, *Auxiliary to Secondary*, *Support*, and *Auxiliary to Support*: to the left are fixed four others, described as the *Tertiary*, *Support of Tertiary*, *Auxiliary of Tertiary* and a *Decayed-leaf*: and along the centre, round the stem of the *Pricipal*, are placed four more, named in their order from above, *Support of Principal*, *Side-piece*, *Trunk-piece*, and *Auxiliary to Trunk-piece*.

All the stems of such a composition are closely united in a single line at the base for a distance of several inches above the surface of the water, and the leaves in most cases overlap one another considerably, only a few of the important ones revealing as much as two thirds of their length. The distribution and balance of leaf surfaces receive considerable attention, so much so, that each leaf, in addition to its other names, is distinguished by the term *Front-surface-leaf* or *Back-surface-leaf*. All leaves are curled or twisted in some way, to show a portion of both sides, but rarely in equal degrees, so that a *Front-surface-leaf* would reveal only a point or edge of its back surface. In Plate L.A. this balance of surfaces is clearly indicated by shading. It is said that in a composition consisting of five leaves, three, including the *End-leaf* and *Finishing-leaf*, should be *Front-surface* leaves, but it appears that no strict rules are followed in this respect, a judicious balance and pleasing variety being alone sought.

Plate XLIV. is instructive as showing defective and corrected arrangements of the Leaf orchid side by side, in which not only the lines but the surfaces of the leaves are altered. An elaborate design, with the same plant in a hexagonal bronze vase, is illustrated in Plate IX.A.

The Leaf-orchid is occasionally used in combination with other trees or plants, sometimes as the auxiliary, and sometimes as the principal member of such double compositions. It is to be seen occupying a subsidiary position combined with the *Nandina domestica*, *Forsythia suspensa*, and large chrysanthemum, and with the small chrysanthemum and *Papaver rhœas* it holds the principal position.

ARRANGEMENT OF MAPLE BRANCHES.

The maple, next to the pine, is the most important flowerless tree used in Japanese compositions. Of this tree, there are two kinds,—the Spring maple, which is red when the young leaves open, and the Autumn maple, which is green in Summer, and turns crimson later on in the year. Floral artists follow several fancy styles of arrangement with maple branches, which are as follows :—

The *Sunrise-arrangement*, in which the leaves of the *Principal* branch should display their front surfaces. The *Sunset-arrangement*, in which the leaves of the *Principal* branch should have their under-sides turned to the spectator The *Cloudy-weather-style*, in which leaves should be curled and sprinkled with spray. The idea of this last style of composition is taken from the appearance of the wild Maples of Ogura-yama near Kioto, the leaves of which are often curled by frost The *Tsuten-arrangement*, in which green leaves are used in the upper, and red leaves in the lower part of the composition. The name refers to a spot called Tsuten, famous for its maple trees, the leaves of which redden from below. The *Tatsuta-arrangement*, so called from a place called Tatsuta, where fine maple trees line the banks of the stream. In disposing the *Principal* mass in this composition, several of the larger leaves should be removed and placed floating in the water of the flower vase, to suggest the leaves which fall off into the river.

A combination of maple branches with chrysanthemums is shown in Plate XXV.A. The maple is often used in water-basins, and sometimes in combination with the iris or other water-plants.

MISCELLANEOUS.

The above description of special arrangements with certain flowers includes those most often introduced into Japanese floral compositions. Nearly every tree and plant, however, the blossoms or foliage of which possess any beauty or attraction, may be seen occasionally introduced into designs, either singly or in combination. The manner of treatment and combination is based upon the principles already expounded as to characteristics of growth, locality, sex, and season, controlled in many cases by traditional fancies.

The examples of compositions with leaf-orchids, *Rhodea japonica*, lotuses, and *Nuphar japonicum*, may be taken as models for arranging most plants having large oval leaves; the designs of irises and narcissus may be followed in employing plants having long, blade-like leaves; the arrangements of clematis and wistaria illustrate the manner of treating trees and plants of the creeper variety; the compositions with plum, cherry, peach, and willow branches serve as examples for the disposition of other straight-branched and blossom-clad trees; and the treatment of chrysanthemums and peonies can be taken as a guide for arranging most plants having large ponderous blossoms.

It only remains to allude to a few examples among the accompaying illustrations which have received no special notice in other parts of the work.

Plate LXV.A. shows an example of the *Rosa indica*, (Choshun), arranged as an auxiliary in combination with branches of peach blossoms in a large flower basket.

Plate XLVII.B. illustrates the *Ilex sieboldi*, (Ume-modoke), in a triple arrangement, placed in a high bamboo vase.

Plate LII.A. exhibits the *Tecoma grandiflora*, (Nozenkazura), in a suspended crescent-shaped vessel, balanced by a separate design of *Calendula officinalis*, (Kinsenkwa), in a standing vessel.

Plate XLVIII.B. shows the cabbage-plant, (Ha-botan), arranged in a globular standing vase.

Plate LXI.B. illustrates the *Dianthus superbus*, (Nadeshiko), placed in a bronze vase.

Plate LXIII.A. represents the *Funkia ovata*, (Giboshi), a large-leaved water-plant, in an arrangement of seven leaves.

Other plants occur in plates which are intended mainly to illustrate different receptacles for flowers. Among these may be mentioned:—the purple magnolia, arranged in a bamboo *Roofed-boat*, Plate XXXIII., the *Patrinia scabiosæfolia*, in a wooden tub, called the *Long-boat*, Plate XXXIII., and again in a hanging vessel on a *Flower-horse*, in Plate XXXVII., the vine, shown combined with small chrysanthemums, in a crescent-shaped vessel attached to a *Flower-horse*, in Plate XXXVII., ivy with camellias, in an inverted bronze umbrella, in Fig. 14 page 70, and the *Aster tartaricus* paired with an arrangement of irises in Plate LVII.

PLATE LXVI

Patrinia scabiosaefolia.

A

PINE, BAMBOO AND PLUM.

B

ORCHID AND CHRYSANTHEMUMS.

C

PLATE LXVII.

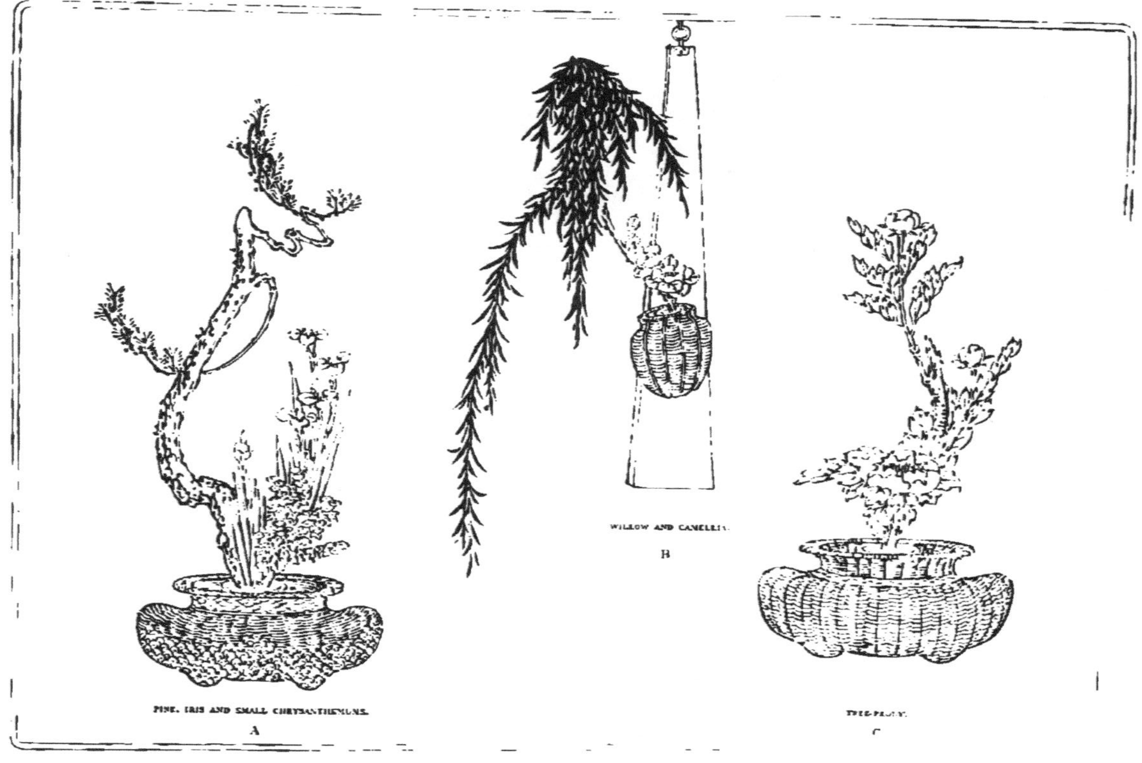

PINE, IRIS AND SMALL CHRYSANTHEMUMS.

A

WILLOW AND CAMELLIA.

B

TREE-PEONY.

C

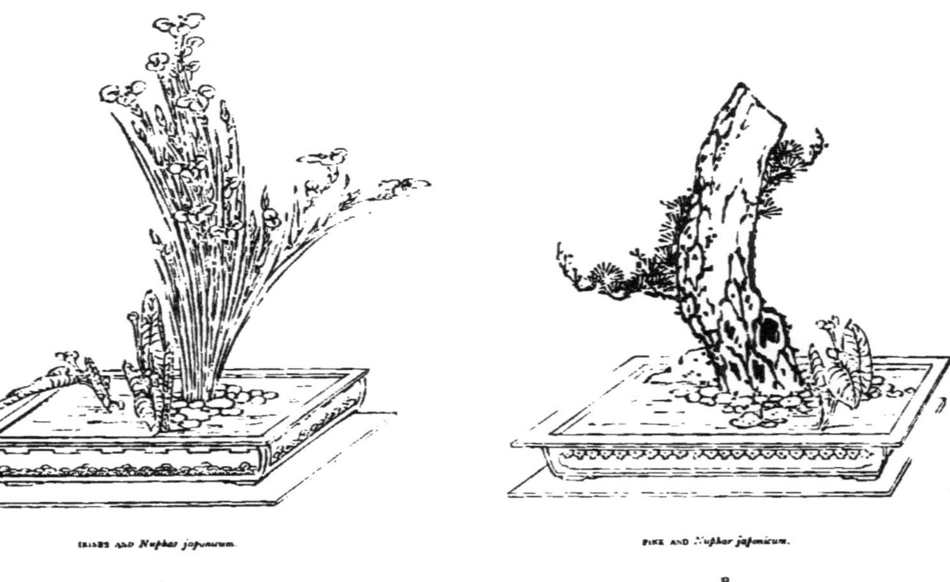

IRISES AND *Nuphar japonicum.*

A

PINE AND *Nuphar japonicum.*

B

WISTARIA, PEONY, IRIS, AND *Nuphar japonicum*.

A

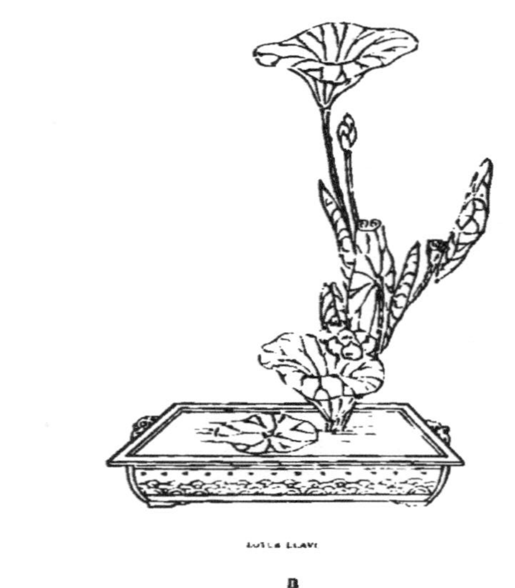

LOTUS LEAVES

B

APPENDIX.

RIKKWA STYLE.

BRIEF reference has already been made to a primitive method of arranging flowers called the *Rikkwa* or *Shin-no-hana* style. In Plate XI. is a diagram showing the theoretical distribution of the seven governing lines in a *Rikkwa* composition. *Shin*, here meaning core or centre, refers to the central and vertical line or mass. *Seishin* is the name given to a smaller mass just below the *Shin*, and in the same line with it. *Soye*, meaning adjunct, is the principal lateral member on the left; *Uke*, meaning dependent, is the most important of the lateral members to the right. Above the *Uke* is the *Mikoshi*, meaning distance, and below it is the *Nagashi*, meaning streamer. The *Mayeoki*, meaning front-piece, is placed centrally at the bottom of the composition. Another line or mass called the *Do-ukuri*, meaning trunk-piece, is often added to these on the left, thus making seven lines in addition to the *Shin*, or central member. Still an additional member called the *Hikaye*, or support, is occasionally introduced, but it is not considered desirable in the most correct compositions.

The SHIN is the principal line of a floral composition. Theoretically it should be central and perfectly vertical, but in the less elaborate styles it is often much bent and diverted. It is said to hold the same relation to the six or seven other lines of a *Rikkwa* arrangement that a lord does to his vassals. The former, therefore, should exhibit stateliness and repose, whilst the latter express force and movement. The auxiliary members of a composition are necessary to its harmony and completeness, but like the various instrumentalists accompanying a graceful dancer, they are administrative chiefly to the central object. By such analogies as these, do writers explain the relative values of the different parts of a *Rikkwa* design.

Material of a thick and heavy nature, either in stem or foliage, should be avoided for the *Shin*. Its characteristics should be straightness, height, and lightness. A branch of young pine is frequently chosen, on account of its erect character and pyramidal termination, which makes a suitable apex to the composition. The pine, moreover, being regarded as the king of evergreens, has a symbolical value which renders it particularly adapted to ceremonial decorations. Anciently this tree alone was selected for the central member or *Shin*. Aftewards, other material became occasionally substituted, preference being always given, however, to trees or plants of an erect and attenuated growth. The bamboo, willow, nandina, fir, oak (*Quercus dentata*), plum, juniper, persimmon, and eularia, are all sometimes employed. The material of the *Shin* to some extent controls the selection of material for the other parts of the design. The wistaria, willow, plum, nandina, and certain other growths, must, if used in the *Shin*, be repeated or *echoed* in some of the secondary lines. Mostly different species of the same genus are preferred for such repetitions.

If, for example, the mountain willow (*Salix paviforia*) forms the *Shin*, the river willow (*Salix purpurea*) is placed at some other point in the design.

When the wistaria is used, an old stump of some hardy tree must be placed in conjunction with it. In all cases where thick branches of semi-decayed or lichen-covered trees are employed for the *Shin*, it is much reduced in height, otherwise it has a top-heavy appearance.

A perfect verticality of the *Shin* is only maintained in the most correct designs. In the less formal arrangements this member is more or less bent over to one side. It is, however, only in the roughest styles that it is allowed to project beyond the edge of the flower vessel; and even in free arrangements considerable care is devoted to the powerful and vigorous posing of this central feature, which must never appear weak or unstable. No lateral curvature must occur until it reaches a point a few inches above the surface of the water from which it springs. The *Shin* is the first member of a floral group to be fixed, and its effect must be carefully studied before proceeding to attach other branches.

The SEISHIN is placed centrally in the composition, just below the *Shin* from whence it derives its name of auxiliary or small *Shin*. It is sometimes arranged so as to hide a portion of the stem of the *Shin*, and is therefore called the *Shin*-concealer. In cases when the *Shin* is bent, the *Seishin* maintains its verticality and marks the central line of the composition. Theoretically the top of this member should be about half way between the bottom and the apex of the composition, though this varies in practice according to the kind of material employed. Some plant, grass, or young tree of erect growth, not too full or leafy at the top, is generally selected, as its function is merely to hide the bareness of the *Shin*. With the diverted *Shin*, however, a leafier material may be used for the *Seishin*, for then it occupies the central gap left in the composition. A certain correspondence in growth must be preserved between these two members Should a branch of young pine be selected for the *Shin*, the *Seishin* should be a spray of some other kind of pine, or if a plant be chosen for the former, the latter must be a plant of somewhat similar character.

The SOYE is the highest member placed laterally in a composition. Anciently it was called the *Tsuyu-uke*, meaning *Dew-receiver*, on account of its form arching at the side of the central member like a branch weighted with snow or dew. It has its visible origin at a point about three inches below the bend of the *Shin* (if the *Shin* be bent), and should be about equal in length to the portion of the *Shin* above the junction. It should contrast in character with the *Shin*. If, as is usual in the most correct designs, the latter is straight and powerful in line, the *Soye* should be of soft and pliant form; but if, on the other hand, the *Shin* is much bent, stiffer material must be selected for the *Soye*. The same kind of contrast is observed in the nature of the foliage;—when that of the one is bare and open, that of the other should be full and leafy.

Following the above principles, the *Shin* having been first placed in position, a branch contrasting with it in character is chosen for the *Soye*, and this is adjusted with as little labour as possible. Much manipulation is apt to produce a weak and artificial effect. The ibuki

(*Juniperus chinensis*) is very suitable for this purpose as its sprays arch sideways, and, though thin and light, are of powerful line. The contour of the *Soye* being of more or less arched character, it follows that when plants, such as the chrysanthemum, having heavy blossoms are used, the flowers tend to hang face downwards: this is considered most objectionable, and several devices are employed to keep the blossoms turned upwards. In speaking of right and left in a *Rikkwa* composition, the terms are applied as if the vase of flowers were a person facing the spectator, and are therefore the reverse of the spectator's own right and left. The *Soye* is generally placed on what is called the right side of the *Shin*, appearing on the left side of the illustration.

The UKE serves as the principal, though not the highest lateral, member on the side opposite to the *Soye*. Whereas the *Soye* is supposed to contrast with the *Shin*, the *Uke*, on the contrary, should accord in character with it. It follows, therefore, that the *Uke* contrasts with the *Soye*. When the *Shin* is large and heavy, the *Uke* should also have similar nature. A stiff, straight, and powerfull *Soye* calls for a bent and pliant *Uke*; also, if the former be lengthy, the latter should be comparatively short. The position of the *Uke* is somewhat lower than the *Soye*, and the two must never be exactly opposite with a bent *Shin*, the *Uke* should spring from a point half way between the top of the vase and the divergence of the *Shin*, and it should curve in a direction contrary to the curve of the *Shin* Some growth contrasting with the material of the *Soye* must be used for this member. If the latter be a sprig of willow, a branch of plum may be selected for the *Uke*. In flowers arranged for religious purposes, the *Uke* branch has a mysterious meaning and is called the *Eko-no-eda* or *Tamuke-no-eda*, according as the ritual is Buddhist or Shintô. In such cases, this branch is made to point in the direction of the relic or image before which the floral design is placed.

The MIKOSHI is the fourth line of the composition. The name is difficult to translate intelligibly by any single word. The meaning is "seen beyond," and the term is frequently applied to objects in a landscape which suggest distance, such as trees viewed beyond a hill or at the limit of the horizon. Bearing in mind that a *Rikkwa* composition is supposed to represent a landscape, the *Mikoshi* branch may be taken as some form in the far-off prospect. On this account it is necessary to avoid using for this member branches of young trees, or tall plants, as they destroy the desired perspective effect, and produce an impression of proximity. It has its proper relation to other lines in the design, being especially the consort of the *Mayeoki*, to be described later, with which it should agree in character—soft or rigid, straight or crooked, light or heavy. On the other hand, it should contrast with the *Seishin*, near which it is placed. The *Mikoshi* generally springs from a point below the origin of the *Seishin*, is of considerable length, and, though arranged more or less centrally, has a slight lateral bend in the direction of the *Uke*. It terminates sometimes above and sometimes below the level of the *Seishin* Occasionally it is called the guardian to the back of the *Shin*, because it is the most important of those branches which spring form the back of the composition.

The NAGASHI is the lowest lateral branch in a composition. It spreads to one side in a sweeping form, arching, dipping, and again rising a little at the extremity. The end must never droop, and it should come somewhat forward in the composition. The *Nagashi* springs from a point about three inches above the mouth of the vase, being about half way between the latter and the origin of the *Uke*, and must always have a slight stilt before its bowed form commences, so as to produce a powerful line expressive of vigorous growth. It ought to be the lowest branch of the composition, and the longest of the lower members. Being placed on the same side as the *Uke*, it is important that the two should not terminate exactly one vertically above the other. The two should also differ in kind of material. In rare cases the *Nagashi* occupies the opposite side together with the *Soye*, and then it is made to contrast with the latter, instead of with the *Uke*. Any growth adapted to the long sweeping form required may be used for the *Nagashi*. In arrangements of pine branches this particular line is called the "incense-burner branch," or the "smoke receiver," names having reference to the low ornamental censer which on ceremonial occasions is placed on one side of the vase of flowers.

The MAYEOKI occupies a front position low down in the design, and has the nature of a bunch rather than that of a line. Formerly, the term included the *Dozukuri* (now a distinct member), making only seven radical parts, inclusive of the *Shin*. For the *Mayeoki*, short bunchy material with a tendency to spread forward must be selected. It should not, however, be of too leafy a nature. The *Rhodea japonica* is often employed. The iris, narcissus, magnolia, and cypress, though frequenty used in other parts of a *Rikkwa* arrangement, are not considered suitable for the *Mayeoki*, and the funkia and nuphar are absolutely prohibited because wanting in rigidity. The character of this member, which may be likened to the bow or rosette binding together the base of the floral group, varies more than any other, according to the rough or finished style of composition. In the most formal style it should be quite central and jut forward without any bend on either side. A lateral bend as it projects forward is allowed in the Gio style, and in the So, or sketchiest treatment, it may be quite irregular, even bending away from the spectator. Arranged in a composition having a bent central line, the *Mayeoki* must conform in character with the *Dozukuri*, spreading right, left, and forward, and creating in combination with the latter member, a hollow called the grotto. Flowers and leaves of attractive shape and colour are employed to produce a bower-like appearance. Ordinary rules, however, require the *Mayeoki* to contrast with the *Dozukuri*: if the one is soft or delicate, the other should be rigid or coarse in nature. Much stress is laid upon the importance of skilfully arranging the *Mayeoki*, which binds the whole mass together. If clumsily placed, the whole design loses unity and repose. It must on no account hide the mouth of the flower vase. Its function is often fulfilled by large leaves, as those of the oak or persimmon tree, which are then subject to all the intricate rules regarding the use of Oba (large leaves),—their exact number, exposure of different surfaces, *et cetera*. Different kinds of Oba must not to be mixed in the same composition. The pine, rhodea, and pteris, are sometimes called the tree *Mayeoki*.

The DOZUKURI, as stated above, was originally one with the *Mayeoki*, and only became distinct in later times. It occupies a central position, assisting to give fullness to the mass. Generally round and full in form, it is nevertheless adapted to the character of the complete composition, serving to correct or tone down any predominating quality. Thus, should the design appear full and leafy, the *Dozukuri* must be somewhat bare. Flowers alone or merely bunches of leaves may be employed. A *Dozukuri* consisting of red blossoms is often added to a floral arrangement of white flowers. It should never screen or hide the water in the vase.

The HIKAYE is an extra branch seldom introduced in the most correct and formal designs. When used, as is sometimes the case in the rougher styles, it is added on the same side as the *Soye*, and below it. To prevent monotony it must be either longer or shorter than the *Soye*, and its distance from the latter depends upon the shape of space or opening which the eye finds desirable. For the trained artist concerns himself as much with the forms of the spaces or hollows in a composition as with the lines and masses, just as the skilled caligraphist studies the hollows of an ideograph rather than the touches by which it is constructed. The *Hikaye* should never be on the same level as the *Nagashi*, which occupies the opposite side of the floral arrangement, nor should it resemble the latter in shape.

The above members may be said figuratively to represent the skeleton, flesh, and simple clothing of a *Rikkwa* composition, but, to complete it, other embellishments are required. For these finishing touches, as well as for certain qualities of perfection aimed at, there are numerous technical terms, which have rather an abstract than a concrete significance, and need not all be enumerated here. It has always been the Japanese art professor's practice to envelope his teaching in mystery by the employment of an immense nomenclature.

Anciently all additions of foliage or flowers to compositions, beyond the seven members already described, were called *ashirae*, or decoration. The Japanese word for gloss or polish is used in a similar sense, having reference to the attachment of blossoms or leafy material to sparse and scraggy lines in order to give body and beauty to the whole. Or, such extra material may be applied in a corrective manner to remove monotony or tone down any defects in the design. Should the bouquet appear too close and heavy, thin and open growth will be added. *Fence-work* is another favourite expression denoting the filling in of a floral composition with secondary material in such a way as to show up its principal features, just as an ornament may be displayed before a screen, or garden objects against a fence. An inexperienced hand is liable to overdo this treatment, feeling dissatisfied with simple masses and openings in parts of his design; but a skilfull designer knows the value of breadth and space, and by their judicious combination will express scenery effects. *Fence-work* may be applied to the back or front of a composition. In the former case, it generally consists of a few plants placed close to the central stem of the *Shiu* and behind the *Dozukuri*. Applied to the front, it consists of material added to the right and left of the *Dozukuri*, and in front of the *Uke* and *Hikaye*.

The terms *Valley* and *Grotto* are used in reference to certain hollows purposely left between the foliage of a design, and intended to be suggestive of landscape effects. The *Valley* is generally situated behind the *Dozukuri*, and in front of the *Seishiu*, and is produced by a skilfull arrangement of foliage or flowers so as to convey the idea of a shady scene. The *Grotto* occupies a place between the *Nagashi* and *Mayeoki*, and is formed by arranging leaves or blossoms arching over a cavity. Much care and attention is given to what is called the *Mizugiwa* or *Water-limit*, a term applied to the base of a *Rikkwa* arrangement with special reference to the manner in which it springs from the base. This springing should always be firm and united, to convey the appearance of vitality and growth, and the bending of different branches must commence from some little distance above. This straight connection or stilt which exists below the curvature of the various members, is made higher in Summer than in Winter compositions, because in the former it is considered refreshing to display as distinctly as possible the water in the vase.

Three modes of arranging the *Mizugiwa* are practiced. In the first, the united stems of the principal members of the composition are left bare at the base; in the second, a small amount of plant or tree cuttings are added; and in the third treatment, the bottom is made as gay as possible by attaching leaves and flowers. The last method, though attractive in appearance and therefore somewhat popular, is condemned by the best masters. Some *Rikkwa* arrangements affect the employment of semi-decayed stumps and branches, in such cases care must be taken that the *Mizugiwa*, or base, is free from decay, otherwise the idea of vitality and growth is lost. In using bamboo stems the distance of the first knot above the water is important for similar reasons.

The terms *Iki* (power, or spirit) and *Utsuri* (reflection) are frequently used conjointly to describe the force of line in one floral member and the reciprocal vigour required in another part of the composition. A powerful *Shin* is said to have *Iki*, and a proper echo or balance of this character in the *Uke* is called *Utsuri*. The word *Iro* in its application to the floral art has the meaning of natural harmony. When the material employed preserves throughout its natural character, with no sign of artificial treatment, branches of straight fibre being used vertically, and branches naturally bent and sweeping disposed laterally, then the composition is said to have *Iro*. Moreover, if there is perfect concord in character and proportion between the bouquet and the receptacle in which it is placed, the same term is employed to signify the harmony produced.

Rikkwa compositions may be broadly divided into two classes, those having a straight and those having a bent *Shin*. These two classes have each their these manners or degrees of elaboration, named respectively *Shin*, *Gio*, and *So*. Plates XII., XIII., XIV., illustrate the *Shin*, *Gio*, and *So* degrees of the straight *Shin* style. The first and more formal arrangement (Plate XII.), is used at marriage ceremonies. It was originally designed to go with the three sacred utensils of a Buddhist altar—the pair of

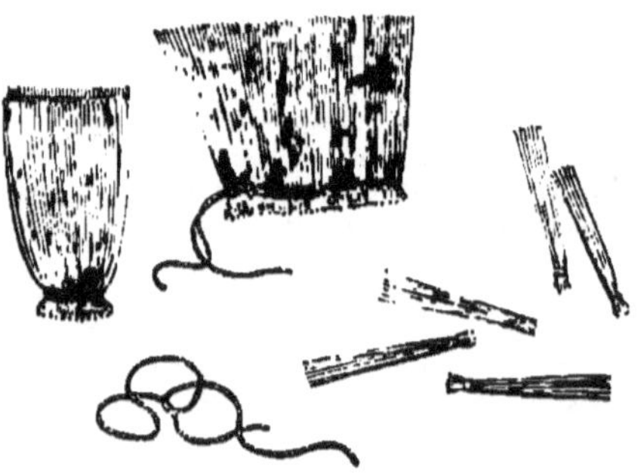

candlesticks and the incense burner. Fasteners for the extremities of cuttings arranged in the *Rikkwa* style, generally consist of small bundles of straw or stubble tied together, and placed in the base.

Plates XV., XVI., and XVII. illustrate the three degrees of the bent *Shiu* style. The principal faults to be guarded against in arranging flowers in the *Rikkwa* style are shown in a figured diagram of Plate XXI. The errors illustrated are as follow:—A, B, Two blossoms of the same kind in a line: C, two lateral branches of the same length side by side, D, cross formed by two branches, E, a branch drooping from the water line over the vase; F, a branch or spray coming forward exactly centrally; G, a large blossom close to the water surface; H, a branch from behind bending round to the front, K, placing other material just below a blossom so as to hide its natural leaves; M, using the flowers only of a plant which possesses fine leaves; L, a branch dipping into the water; J, a branch twisting back towards the wall, I, one branch crossing another diagonally.

The *Rikkwa* Stump arrangements are illustrated in Plates XVIII., XIX., and XX. For such floral compositions a low broad basin or bowl is used, and instead of the ordinary stubble fastening, a wooden framework is let into the basin to hold steady the heavy stubs and branches used. This framework is afterwards to a great extent concealed by the sand, pebbles, and water. The Sand-bowl arrangements of flowers, as these are sometimes called, are also controlled by a theory of seven governing lines. Whereas in the *Shiu-no-hana* or Standing-vase arrangements, the tendency is vertical; in the Sand-bowl arrangements, the tendency is horizontal. The high becomes low, and the narrow broad. In the more finished style, one thick stump is employed in continuation with other materials, (see Plate XVIII.). Here we have a plum stump with branches, combined with pine, camellia, and narcissus.

Plate XIX. illustrates the double stump arrangement which is followed in the rougher style. The idea of sex is applied to such arrangements, the stump to the right being called *male*, and that to the left *female*. In this composition various kinds of pine, irises, reeds, and bamboo grass are employed.

Plate XX illustrates a combined *Rikkwa* and Sand-bowl arrangement, suitable for placing upon the ornamental shelves of a chamber. The upper arrangement is on the principle of the *Shin-no-hana* or erect style, but is somewhat lowered and broadened out on account of its high position. The lower arrangement is in the rough Sand-bowl style, with double stump of pine, irises, bamboo, and ferns.

IKENOBO STYLE.

The *Ikenobo style* of flower arrangement has of late years become somewhat popular among amateurs in Japan. It is the revival of an ancient method, to be traced directly to a modification of the original *Rikkwa style*, to which certain resemblances may be observed. In Plate LXVI. at B is shown an *Ikebono* design with pine, plum, and bamboo, which it is interesting to compare with the *Rikkwa* designs in Plates XI. and XII.

The *Ikebono* school, however, does not adopt an abundance of different material in one composition as in the *Rikkwa style*, but limits itself to the combination of two or three growths. The most striking characteristic of this method is the close and bunchy nature of the designs as compared to those of the Enshiu and other schools The lineal character, which is so marked in the latter, to a great extent disappears, though the outline of the floral arrangements produced contains a suggestion of the radical lines. The triangular contour always predominates, and this may be seen very distinctly in Plate LXVI. A. and C., Plate LXVIII. A., and Plate LXIX. B.

In effect, the arrangements of the *Ikebono style* appeal to the European taste as more natural and less conventional than those of the Enshiu style. They are, however, equally subject to elaborate rules not unlike those of the rival schools. The following illustrations from Plate LXVI. to Plate LXIX. give examples of *Ikenobo* arrangements, in vases, baskets, and sand-basins.

印刷所　東京市牛込區市ケ谷加賀町壹丁目十二番地　株式會社 秀英舍工場

印刷者　東京市京橋區西紺屋町二十六番地　株式會社秀英舍内　青木弘

發行者　東京市京橋區西紺屋町十七番地　東京府士族　高橋道保

著作者　東京市京橋區西紺屋町十三番地　英國人　ジェー、コンダー

明治三十二年六月五日訂正再版發行

明治三十二年六月一日訂正再版印刷

明治二十四年六月十日印刷并出版

www.ingramcontent.com/pod-product-compliance
Lightning Source LLC
Chambersburg PA
CBHW032048230426
43672CB00009B/1516

* 9 7 8 3 3 3 7 1 7 4 9 2 7 *